GW00707670

OUTDOOR PIGS

Principles and Practice

Proceedings of a conference held at Oxford University
3 and 4 April 1989

Edited by

B A Stark, D H Machin
and
J M Wilkinson

CHALCOMBE PUBLICATIONS

First published in Great Britain by
Chalcombe Publications, 13 Highwoods Drive,
Marlow Bottom, Marlow, Bucks SL7 3PU

© Chalcombe Publications 1990

ISBN 0 948617 20 9

CONTENTS

CONTRIBUTORS

Sue Corning

ADAS National Pig Specialist,
MAFF, Block B, Government Buildings,
Brooklands Avenue, Cambridge CB2 2DR

Cathy Shepherd

Pig Producer,
Allenford Farms, Damerham,
Fordingbridge, Hampshire

Dr A E Wrathall

Head of Diseases of Breeding Department,
MAFF, Central Veterinary Laboratory,
New Haw, Weybridge, Surrey KT15 3NB

Dr J M Bassett

Lecturer in Endocrinology,
University of Oxford, Growth and
Development Unit, University Field Laboratory,
Wytham, Oxford OX2 8QJ

Dr W H Close

Group Leader Pig Production,
AFRC Institute for Grassland and Animal
Production, Church Lane, Shinfield,
Reading RG2 9AQ

P Poornan

Chief Executive, Lys Mill Ltd,
Watlington, Oxford OX9 5ES

Dr D H Machin

ADAS Regional Pig Specialist,
MAFF, Government Buildings, Marston Road,
New Marston, Oxford OX3 0TP

N M Beynon

Head of Department of Agriculture,
Berkshire College of Agriculture,
Hall Place, Burchetts Green,
Nr. Maidenhead, Berks SL6 6QR

Dr M Bichard

Technical Director,
The Pig Improvement Company Limited,
Fyfield Wick, Abingdon, Oxfordshire OX13 5NA

Prof A J F
Webster

Department of Animal Husbandry,
University of Bristol, Langford House,
Langford, Bristol BS18 7DU

FOREWORD

by J E Riley

Chief Livestock Adviser, Ministry of Agriculture, Fisheries and Food,
Agricultural Development and Advisory Service, Nobel House, 17 Smith
Square, London SW1P 3JR

The recent resurgance of interest in outdoor pig production has arisen for a number of reasons, including the relative profitability of outdoor sows compared with cereals at the present time, animal welfare considerations, high capital demands of intensive pig production, a succession of mild winters and a demand by consumers for "wholesome meat". Successful outdoor pig production demands the right type of soil, a mild climate and, most importantly, innovative management.

There is, of course, nothing new about outdoor sow production, and over the years the system has attracted and benefitted from the efforts of innovators and pioneers. Richard Roadnight was responsible for the development of the Britwell Blue and a labour saving system based on farrowing twice a year, in March and September. Robert Bowden has also had a major influence on outdoor pig production, both in this country and abroad, and he may be credited with another major breakthrough in the mid 1970s, namely weaning at 3 weeks.

During the last few years, one of the major problems faced by the outdoor pig producer has been market resistance to the pigs produced, both weaners and animals for slaughter. Any pig that shows even slight colouring has been penalised in the market place in recent years, which is a complete reversal of the position in the 1960s when the coloured pig attracted the premium and the white pig was penalised. Today we are experiencing another break-through—production by the breeding companies of breeding stock selected specifically for outdoor systems. The aim with these stock is to reduce the chances of a coloured animal and therefore marketing penalties.

My estimate is that 8 to 10 percent of the national breeding herd spend some of their time outdoors. Some people suggest that 30 to 40 percent of the national herd could be run effectively outdoors. However, it must be emphasised that outdoor pig production is not a quick way to get rich. The system has problems, including the control of resources (pigs and capital), the motivation of staff, and autumn infertility, and research into the specific problems of outdoor pigs is urgently needed.

This conference was held to provide a forum for producers with outdoor pig herds, researchers, advisers and members of commercial companies with an interest in outdoor pig production, to consider the varied problems associated with keeping pigs outdoors. Important issues for the industry were discussed, together with the scope for technical developments, and requirements for research.

CHAPTER 1

OUTDOOR PIG PRODUCTION IN THE UK

S Corning

SUMMARY

Pig producers operating outdoor systems and those considering starting an outdoor enterprise have little published information on which to base their decisions, and there has been a lack of research on the specific problems and requirements of outdoor herds. Recording schemes generally relate to a small sample of herds and data must be treated with caution. However, in general, sows in outdoor herds have recorded fewer litters per year compared to indoor herds, but total numbers of piglets produced per sow per year can be similar for the two methods of production.

Outdoor pig production offers an advantage over indoor systems in terms of capital requirements for building and equipment, but this advantage can be considerably reduced if land has to be purchased or rented. Realistic land charges are an important consideration when comparing systems financially. Recording Scheme results suggest that feed, labour and farm transport costs are slightly higher for outdoor systems, but power, water and bedding costs are lower. Outdoor herds also have a lower building charge but this may be partially offset by a pasture charge. Overall, using the limited records available, even in a year of poor profitability for the pig industry outdoor producers have recorded a higher net margin per sow than indoor herds. However, the financial advantage in terms of lower costs for extensive systems could be negated by poor pig performance under adverse environmental conditions and/or poor management.

Improvements in the number of piglets reared per sow per year, piglet survival and sow feed costs can all lead to marked improvements in margins, and research into these areas is likely to be cost-effective. The potential of outdoor systems to produce a premium quality product should also be examined critically. The availability of suitable land for outdoor pig production is likely to be less limiting to the expansion of outdoor herds than the availability of stock people to operate the system. In comparing production systems based on welfare considerations, it is important that all aspects of management and environment are taken into account.

INTRODUCTION

Recent renewed interest in outdoor pig production has increased the need to review past developments in outdoor pig production, to assess the advantages and disadvantages of current systems, and to consider the future needs of the industry and likely developments. Published information is, however, lacking, as illustrated by a recent literature search on outdoor sows. This revealed 20 references, of which 8 were in Russian, and 5 related to Canadian sows kept in climatic conditions which reached −30°C; the remaining papers were of a general nature relating to the UK pig industry.

It must therefore be emphasised that very limited information is available both for people thinking of outdoor production, and for existing outdoor producers wishing to look critically at ways of changing and improving their current system.

SYSTEMS

What is outdoor pig production? What is defined as an outdoor system?

Table 1.1 shows data for a number of Pig Recording herds recorded on the outdoor system within the Meat and Livestock Commission (MLC) scheme. Although the information is from a sample of herds and is thus not necessarily representative of the whole country, it provides a guide to the type and number of outdoor herds which would be expected. In terms of dry sow accommodation, about 6 percent of the MLC herds are defined as having dry sow accommodation outdoors. This compares with 50 percent or more kept in stalls and tethers.

Table 1.1 Dry sow accommodation

	% Herds
Sow yards	34
Tether house	28
Stalls	24
Cubicles	6
Outdoors	6
Others	2

Source: MLC Pig Yearbook, April 1987. Meat and Livestock Commission, Milton Keynes.

Table 1.2 Farrowing accommodation

	% Herds
Fully enclosed purpose-built	54
Non purpose-built (conversion)	32
Combination of above	7
Outdoors	4
Individual purpose-built huts	3

Source: MLC Pig Yearbook, April 1987. Meat and Livestock Commission, Milton Keynes.

Definitions of outdoor sows, in the main, consider herds in which sows farrow outside as being truly representative of the outdoor system. On this basis, it appears that from the same sample of herds about 4 percent fall into the outdoor category (Table 1.2), although some of the individual purpose-built hut systems may also operate to some extent outdoors.

Thus about 5 percent of the breeding herds, which may be about 6 to 8 percent of the sow population, or 50,000 to 60,000 sows, are kept on an outdoor system. There is, however, considerable variation throughout the country. In Oxfordshire 25 percent of sows may be kept on an outdoor system, the percentage in Hampshire may be as high as 40 percent, while East Anglia is another area with considerable outdoor pig production.

CAPITAL REQUIREMENTS OF OUTDOOR SYSTEMS

Capital investment is one of the first issues to be raised when considering outdoor pig production, as this is much lower than for indoor systems. Table 1.3 illustrates capital requirements per sow with data taken from the Cambridge Pig Management Scheme which relates mainly, but not exclusively, to East Anglian herds.

Table 1.3   Capital requirements per sow

	Indoor (£)	Outdoor (£)
Value of sow	122	118
Share of boar value	14	12
Buildings and equipment	415	216
Working capital	111	111
Total capital (excluding land)	662	457

Source: Pig Management Scheme Results for 1988. R.F. Ridgeon, Department of Land Economy, University of Cambridge.

It must be emphasised that the indoor and outdoor capital requirements recorded by the Cambridge Pig Management Scheme in Table 1.3 refer to herds that are already operating in the Recording Scheme; they do not refer to capital requirements for a new site. This is an important point when looking at some of the costs of building and equipment that are quoted.

Evidence suggests that there is very little difference in the capital requirements for stock between indoor and outdoor systems. Table 1.3 shows a slightly higher value for boars on an indoor system, but the opposite could be experienced, taking into account the sow : boar ratio expected with the different systems. Working capital is similar for the two methods of production. However, capital requirements show a significant difference.

The major differences in capital requirements between indoor and outdoor systems are building and equipment costs, which Table 1.3 suggests are approximately double for an indoor system. If the herds were being established on a new site, the difference would be even greater. A significant point, however, is that this comparison of capital costs excludes land, as producers in the Recording Scheme all had land available.

This point is very important for people considering starting an outdoor unit, as the cost of purchasing land could easily be £200 to £300 per sow. Recording schemes often quote a figure of £6 to £10 per sow per annum for the cost of land to run an outdoor system. If land is rented that figure could easily be doubled, and if land is bought the figure may be around £40. Producers considering setting up an outdoor pig enterprise should ensure prepared budgets are realistic; in many situations the cost of land does not appear to be taken fully into account.

PERFORMANCE OF OUTDOOR HERDS

It must be emphasised that very limited information is available on outdoor herds, and the MLC provides one of the largest sources of recorded data. It can be seen from Table 1.4 that herds on outdoor systems are relatively large, with an average herd size of 515 sows in the sample; this is in part a reflection of lower capital costs associated with establishing an extensive system.

Table 1.4 Comparison of the performance of outdoor and indoor herds

	Outdoor	Indoor
Average number of sows and gilts	515	200
Litters per sow per year	2.20	2.26
Pigs born alive per litter	10.24	10.59
Mortality of pigs born alive (%)	10.6	11.2
Pigs reared per sow per year	20.1	21.4
Sow feed (tonnes per sow per year)	1.29	1.15

Source: MLC Yearbook, April 1988. Meat and Livestock Commission, Milton Keynes.

The replacement rates shown in Table 1.5 for 1983 and 1987 are very similar, 39 to 40 percent, which is probably comparable to that for indoor herds.

The number of litters per sow per year, about 2.2 for MLC recorded herds, was very similar for 1983 and 1987. Data from the Cambridge Pig Management Scheme also show that there has been little change in the number of litters per sow per year over the last 4 or 5 years. This may reflect ongoing problems of infertility and/or service management. However, in

Table 1.5 Performance of MLC recorded outdoor herds

	1983	1987	1987 Top third of herds
Number of herds	35	45	15
Average number of sows	395	515	542
Replacement rate (%)	36.7	39.9	43.5
Litters per sow per year	2.19	2.20	2.30
Pigs reared per sow per year	19.3	20.1	22.3

Source: MLC Pig Yearbook, April 1988. Meat and Livestock Commission, Milton Keynes.

terms of pigs reared per sow per year (Table 1.6), over 4 years there has been an increase of about a pig per sow per year. Taking into account that the number of litters has not changed, there must have been more pigs born per sow, or a lower mortality rate or a combination of both, leading to more pigs weaned per sow. It is important to assess the reasons for the improvement in performance, particularly the possible influence of weather conditions on piglet mortality.

Table 1.6 Performance of outdoor and indoor herds recorded in the Cambridge Pig Management Scheme

	1985		1988	
	Outdoor	Indoor	Outdoor	Indoor
Weaners per sow per year	18.6	21.6	21.1	21.9
Margin per sow (£)	76.63	80.35	26.59	−19.93

Source: Pig Management Scheme Results for 1985 and 1988. R.F. Ridgeon, Department of Land Economy, University of Cambridge.

Caution is needed when outdoor producers define production targets, or improvements needed in performance. The absolute target number of piglets per sow per year should be related to profitability. For example, in one situation a herd may produce 21 piglets per sow per year but the level of costs incurred requires that 23 piglets are reared to break even. By contrast, sows in another herd may produce 18 piglets per year, but the producer shows a profit by rearing over 16 piglets per sow per year.

Data from Table 1.6 emphasise this point. For indoor and outdoor systems, the number of weaners per sow per year was respectively 18.6 compared to 21.6 in 1985, and 21.1 compared to 21.9 in 1988. However, 1985 was a reasonable year for pigs in terms of overall profitability and the indoor herds had a margin per sow of about £80, while the margin per sow in the outdoor systems was about £76. Because of the lower break even point, in a profitable year the two types of systems made very similar margins per sow, despite fewer piglets per sow with outdoor systems. By contrast, 1988 was a year of low profitability for pigs in general, but an improvement in the performance of the outdoor herds led to a difference in margin of about £40 per sow in favour of the outdoor herds. An improvement in performance made outdoor pigs more competitive, and even in a year of low profitability they showed a positive margin per sow.

The sample of herds shown in Table 1.7 produced about 2.2 litters per sow per year, which is a similar level to that found in most recording schemes for outdoor herds. This value compares with 2.36 litters per sow per year for indoor systems; thus there is a real difference between the types of systems, which will be considered further in subsequent chapters. As the number of litters per sow per year is a significant financial consideration, it is important to pay attention to improvements in this area.

Table 1.7 Comparison between outdoor and indoor herds selling weaners/stores

	Outdoor	Indoor
Litters per sow per year	2.21	2.36
Live pigs born per litter	11.2	10.6
Weaners per litter	9.6	9.3
Weaners per sow per year	21.1	21.9
Sow feed (tonnes per sow per year)	1.44	1.23
Compound feed (% total feed)	91	70

Source: Pig Management Scheme Results for 1988. R.F. Ridgeon, Department of Land Economy, University of Cambridge.

Table 1.7 also indicates that outdoor herds produced 11.2 piglets born alive per litter and indoor herds 10.6 piglets per litter. Thus outdoor herds may not produce as many litters per year but the number of pigs born alive appears to be satisfactory compared to indoor systems. (Note that MLC data (Table 1.4) show the opposite trend, illustrating that recorded figures must be interpreted with caution). The number of pigs weaned per litter, at 9.6 for outdoor herds and 9.3 for indoor systems (Table 1.7), suggests a pre-weaning mortality on the outdoor system of about 14 percent compared to 12 percent for the indoor system. A higher number of pigs born alive in the outdoor system compensates for the lower number of litters per sow per year, to give a very similar number of weaners produced by both systems.

In terms of annual feed requirements, outdoor sows used 1.44 tonnes compared to 1.23 tonnes per sow indoors (Table 1.7). This is expected as the environment outside can be cold and wet, sows are competing with other animals in most group situations, and there is obviously more wastage even with the large nuts and rolls used outdoors. Based on the figures given in Table 1.7, this difference in feed requirements may be worth about £22 per sow per year.

The additional feed costs may have been accepted until now because producers know that under adverse environmental conditions on outdoor units higher feed levels are required to maintain sow condition and productivity. However, feed costs may be an area which has received insufficient attention because, although more feed may be needed, this does not always have to be purchased at a high price. The use of bulky alternative feeds, as discussed in Chapter 7, may help to reduce the difference in feed costs between indoor and outdoor systems. Table 1.7 shows that about 91 percent of the outdoor herds recorded used purchased compound feed and, while this situation is unlikely to change markedly because the production of biscuits or cobs is difficult with small scale facilities, feed costs should nevertheless be closely scrutinised.

COSTS AND RETURNS

Data from the Cambridge Pig Management Scheme suggest that feed, labour and farm transport costs are all slightly higher for outdoor systems, as shown in Table 1.8. The difference between labour costs of £5.04 and £4.92 per pig for outdoor and indoor systems respectively is smaller than expected, and insignificant in practice. Power and water show the largest difference in 'variable' costs between systems, at 80 pence per sow. Higher power costs would be expected on indoor systems but it is not possible to separate water costs from these figures. Miscellaneous costs were slightly higher for indoor herds, and bedding costs were about double for indoor compared to outdoor herds. AI costs were similar for both systems and, overall, the difference in 'variable' costs was about 66 pence per pig in favour of the outdoor herds.

In terms of the 'fixed' costs incurred, maintenance and equipment costs were similar. Although these costs might be expected to be higher for indoor

Table 1.8   'Variable' costs and returns for outdoor and indoor herds

	Outdoor (£)	Indoor (£)
Feed	16.40	16.05
Labour	5.04	4.92
Farm transport	0.66	0.39
Veterinary	0.40	0.67
Artificial insemination (AI)	0.06	0.08
Power and water	0.19	1.00
Miscellaneous	0.42	0.60
Litter	0.13	0.25
	23.30	23.96

Source: Pig Management Scheme Results for 1988. R.F. Ridgeon, Department of Land Economy, University of Cambridge.

herds, it should be noted that the figures in Table 1.9 relate to outdoor herds finishing weaners, therefore some weaner accommodation costs are recorded in the scheme. Building charges were approximately half for outdoor herds. There was also a pasture charge for outdoor systems of about 66 pence per pig reared, and overall the building and pasture charges amounted to 141 pence for the outdoor systems compared to 162 pence for the indoor systems, which is perhaps not as large a difference as might be expected. Obviously, when producers wean straight from the field, no rearing accommodation is required and the difference between the systems is greater.

Table 1.9 'Fixed' costs and returns per pig for outdoor and indoor herds

	Outdoor (£)	Indoor (£)
Maintenance	0.34	0.57
Equipment	0.29	0.24
Building charge	0.75	1.62
Pasture charge	0.66	—
Other costs	23.30	23.96
Stock depreciation	0.90	1.26
Total costs (excluding interest)	26.24	27.65
Weaner price (net)	27.50	26.74
Margin (excluding interest)	1.26	−0.91

Source: Pig Management Scheme Results for 1988. R.F. Ridgeon, Department of Land Economy, University of Cambridge.

Total costs given in Table 1.9 amounted to £26.24 per pig for outdoor herds and £27.65 for indoor systems. When these costs are deducted from the weaner price the result is a margin of £1.26 per pig for outdoor systems compared to a loss of 91 pence for indoor systems. These figures relate to 1988, which was not a profitable year for the pig industry, but the difference in margin of approximately £2 per pig between the systems is probably realistic. An attraction of outdoor systems is that they appear to produce a positive margin even in a year of poor profitability.

Weaner prices given in Table 1.9 are not strictly comparable between systems because for outdoor systems the price of £27.50 related to a 31 kg pig, while for indoor producers £26.74 related to a 28 kg pig. These data provide no evidence of a premium paid for outdoor weaners, as the price for a heavier pig from the outdoor system was about 7 pence per kg liveweight less than that for a 28 kg pig. The difference was, in fact, larger than expected for this difference in weight.

THE CURRENT POSITION

Overall, the margin over fixed and variable costs given in Table 1.9 was £1.26 per pig for outdoor systems compared to a 91 pence per pig loss for indoor systems. Interest charges were higher on indoor systems, as were capital costs for buildings and equipment, although adding a capital land charge would increase outdoor herd costs. In terms of profitability, after subtracting interest charges neither indoor nor outdoor systems had a profitable year in 1988, although at 21 pigs per sow per year, outdoor herds made about £26 per sow per year compared to a loss of about £20 for indoor systems, with a difference in margin of about £40 per sow. The return on capital in 1988 was about 6 percent for outdoor herds compared to a loss of 3 percent for indoor systems. In a more profitable year, comparable figures may be about 19 to 20 percent return on capital for outdoor herds and 10 to 12 percent for indoor systems.

Thus, in terms of production levels and financial charges outdoor herds appear to have improved their performance. Care must, however, be taken when interpreting previous records and setting targets, to ensure comparisons between the systems are made on the same basis. A better return on outdoor systems cannot be guaranteed, particularly if the cost of land and establishing the system is taken into account.

Summarising some of the financial differences between outdoor and indoor herds, it is suggested that outdoor pig producers would expect to pay about £20 more per sow per year for feed, and to save about £20 per sow per year on variable costs. In addition, fixed costs might be reduced by £25 to £50 per sow per year and, overall, outdoor systems might reasonably expect to save £25 per sow per year. Table 1.10 shows the decrease in level of performance compared to indoor herds which could be tolerated by outdoor systems before this £25 advantage is nullified. Clearly, with adverse environmental conditions and a low standard of management, poor herd performance could negate the initial financial advantages of outdoor systems.

Table 1.10 Change in performance of outdoor herds compared to indoor herds to nullify the typical cost saving of £25 for outdoor herds

	Change for outdoor sows (£/sow/annum)
Feed costs	+20
Other variable costs	−20
Fixed costs	−25 to −50

Source: ADAS, private communication.

AREAS FOR IMPROVEMENT

Despite outdoor pig production being regarded as a low cost system, it is still important to examine critically the areas where improvements would be most advantageous. As an illustration, it is possible to consider the financial implications of changes in physical performance in a 360 sow herd producing weaners. An improvement of 1.5 piglets per sow per year could be worth an extra £10,000 in margin; this is thus an area where improvements are financially attractive. If piglet survival increased by 5 percent, this would be worth about £9,000 and, if sow feed costs were reduced by £10 per tonne, margins could be improved by about £4,000. Improvements in other parameters such as the price of pigs, finishing feed conversion ratio, and number of pigs born alive per litter would also have effects on profitability.

This example is based on one set of data. Calculations should be done on individual herds to highlight priority areas for improvements.

Areas where investment is justified need to be considered. A decrease in mortality of 1 percent would improve margins by about £1300, which in present capital value terms is an investment of about £5000. Table 1.7 shows mortality figures of about 14 percent on outdoor systems, but the reasons for this relatively high mortality are largely unknown. Many producers find that their ark systems operate with few problems, but little research has been carried out to study environmental conditions within the arks and to determine if improvements could or should be made.

A short ADAS investigation recorded temperatures in insulated and uninsulated huts. With outside temperatures from 12°C down to −6°C, the temperature in the uninsulated huts was −5°C for between 13 and 30 percent of the time, whereas in an insulated hut it was over 12°C for 100 percent of the time. A reduction of 1 percent in piglet mortality would justify an extra £100 on the cost of every farrowing hut on a 200 sow unit! All key production areas should be examined critically to decide where investment can be justified.

Other areas which merit further attention include systems for bulky feeds, such as fodder beet, possibly in conjunction with electronic sow feeders on outdoor systems. Silage, particularly as large bales in a self-feed hopper, may be a useful bulky feed for sows. There is also scope for looking more critically at weaner quality, and at systems for finishing outdoor pigs to produce a premium product.

WHY CONSIDER OUTDOOR PIGS?

In some instances, outdoor pig production is considered when spare land is available. Based on the number of days that land is at field capacity, which relates soil type to rainfall, it is possible to indicate areas of the UK with land well suited to outdoor pig production. Using these criteria, the availability of suitable land will not be the limiting factor for people wishing to go into outdoor production.

A greater problem will be the number of stock people available to keep outdoor pigs effectively. An estimated 20,000 people are engaged in pig production and about 650 people entering the industry each year are needed, assuming that people stay for about 30 years. On a 25 year basis 800 people per year would be needed. Stockmen could be a limiting factor for pig production, in particular with outdoor systems, which can be more demanding and may result in a higher staff turnover rate.

Outdoor sows are often suggested as a favourable welfare option. Three key welfare issues which have been considered in pig production are age at weaning, dry sow accommodation and the use of slats without bedding, particularly for finishing pigs. Of these points, outdoor pig production would have most influence on dry sow housing and this is an area that should be examined.

A personal appraisal of welfare points comparing outdoor and indoor systems indicates that the degree of control over parameters such as feed intake, bullying, observations and working conditions varies considerably between the extremes of stalls and paddocks, and will inevitably be affected by the standard of management.

The Welfare Codes recommend alternatives to stall and tethers, with outdoor systems as one option. However, it is important that research and development information is available to ensure that all production systems can be fully evaluated.

DISCUSSION

It is often assumed that extensive systems will be associated with more problems in winter than in summer, but this is not always the case. Putting intensively bred pigs outside can lead to considerable problems in summer in terms of sunstroke and sunburn. On the other hand, stock people tend to have more problems in winter conditions than during summer weather.

Interpretation of sow behaviour in terms of welfare is very difficult, and there is little definitive evidence to compare production systems. Some degree of bullying has been recorded in most grouped sows and may be natural behaviour. The pig industry has responded to welfare pressures but it is important that all alternative production systems are fully evaluated.

CHAPTER 2

KEEPING PIGS OUTDOORS: A PRODUCER'S VIEW

C Shepherd

SUMMARY

There is much variation in management and equipment between different outdoor pig units, and a system which operates successfully on one unit may not necessarily produce good results on another. This paper is based on the personal experience of a producer in an area moderately suited to outdoor pig production.

The first factors to consider with an outdoor pig unit are climate, soil type, topography and geographical location. Stocking rate may be influenced more by the integration of pigs within an arable rotation than by theoretical considerations. Typical outdoor sows are Landrace cross Saddleback, but these sows are being replaced by other crosses. The possibility of having to finish pigs at a heavier weight in the 1990s will influence the type of both sow and boar. Some outdoor units have changed to hybrid boars while others remain committed to the Large White. There are both advantages and disadvantages to breeding replacement gilts compared to buying stock.

Outdoor sows are managed as a group, not as individuals, and variability in body condition within a group must be minimised. Management, particularly feeding, at weaning and at service is critical for good conception rates, and many successful variations operate on different units. Little is known about the effect of feeding strategy on the implantation or resorption of embryos in the outdoor situation, and the condition of sows at weaning tends to influence feeding level and the length of time higher rates of feed are given. Sow to boar ratios for outdoor pigs vary from 14 to 20, and the biggest breeding problem is summer or autumn infertility. Subsequent lack of piglets can cause problems in finishing units.

Sows should produce as many live pigs as possible, with the minimum range in age of piglets within a group. Farrowing policy is strongly influenced by the size of the unit. Management has a marked effect on piglet survival, but there are no clear guidelines on many aspects of this. The best feeding system for lactating sows is also unclear; feeding on the ground often increases variability in body condition, and ad lib feeders may result in excessive feed consumption, although this is not inevitable. Constant attention must be paid to pig health but views differ on the need for preventive medicine strategies. A wide range of diseases may be encountered.

Adequate numbers of good staff are essential for successful outdoor pig units. Turnover is often higher on outdoor units and the imminent problem of decreasing numbers of people entering the pig industry is likely to be amplified for outdoor producers. Size of unit and level of productivity are closely related to labour availability and staff motivation.

INTRODUCTION

This paper is based on experience helping my father to run an outdoor pig unit in conjunction with cereal production, close to the Hampshire/Wiltshire border. It concentrates on aspects of the pig unit where there are problems and uncertainties, and where answers are not clear. A point to remember, however, is that what we do on our unit may be totally contradictory to another unit, and within the basic framework of outdoor pig production there is much variability on individual units.

LOCATION OF OUTDOOR PIG UNITS

When considering an outdoor pig unit, the first factors to take into account are climate, soil type, topography and location in the country. Our unit in Hampshire is situated on a light loam, chalky soil, with about 32 inches of rainfall a year. Theoretically we are in an area moderately suited to outdoor pigs and we have typical land. The importance of siting an outdoor unit on suitable land within the farm was clearly demonstrated in 1989 when rain followed a very dry winter; the pig fields quickly became muddy, and with less suitable land this would have been much worse.

A second consideration when locating an outdoor unit is the required stocking rate. Many fields can potentially be used for pigs providing that the stocking rate is adjusted so that ground conditions do not deteriorate too badly. However, on our farm pigs are part of an arable rotation, which means that the pig unit is not always in a position to specify the required stocking rate, because only a certain amount of land is available for the pigs. Whilst the theoretical stocking rate is 15 to 18 sows per hectare (7 to 8 sows per acre), it is questionable whether this is achievable in practice.

This year (1989) we have a farrowing field a mile and a half away from our dry sow paddocks, which causes extra management problems, movement of sows, time, labour, transport, and equipment. Two years ago all sows were on the same site, which was excellent for management of the pigs, but the stocking rate had to be increased and the yield of the subsequent crop was not well-received by the arable side of the farm.

We can also still see trackways in cereal crops where there were pig paddocks several years ago, and I question whether outdoor pigs necessarily make a useful contribution to the arable rotation in all situations. I wonder whether we should consider adapting the arable crops around the pig rotation, rather than adapting the pigs to fit in with the arable rotation as traditionally done. The effect of the pigs on subsequent crops must also be considered. Apart from a "puddling" effect on soil and its subsequent influence on yield, mayweed and couch problems frequently follow pigs and these weeds then have to be treated in the subsequent arable crop.

TYPE OF STOCK

Having considered location of the pig unit and the pig rotation, the next issue is type of stock to use. We have always produced a typical Landrace cross Saddleback sow, since the first ones were bought in the mid 1960s. This type of sow has served us very well over the years, and we have had no complaints. However, there is a tendancy within the industry now to think ahead and to consider that a different type of animal is needed, and we think that we should replace some Saddleback cross Landrace with alternative crosses. At the same time, however, on the productivity side, we still have further to go with this sow because, although in total about 11.5 piglets are born per litter, we have an average of 0.75 piglets born dead, and we should be able to increase the percentage born alive. We also know that this sow is very docile and we know how to manage her.

The other aspect of sow type is the need to produce a finished pig which grades well, and which can perhaps be taken through to a heavier weight in the 1990s. At the moment we do not have a problem as our pigs grade satisfactorily for light pork. However, by the 1990s we may have to take pigs to heavier weights and our stock may not grade. But will there be no niche within the market for a lighter pig, and will we all have to produce heavier pigs? I suggest that producing heavier pigs will cause many outdoor pig producers problems because of lack of buildings to house bigger animals for a longer period.

The type of boar put to the sow affects the source generation, and we questioned whether we should retain the pure bred Large White or change to the hybrid type boars. We have moved totally to the HY boar because we like the conformation of the pigs on the source side and the boars work well. But are we right? Other units are very satisfied with Large Whites and are adamant that they will retain them.

Whilst we are committed to Landrace cross Saddleback sows in our particular circumstances, we are also running some other lines to see if they perform better. We intend to observe their productivity and finishing performance before we make a decision to change. The animal which suits one farm does not always suit another.

Another question is whether we should be breeding our own gilts or buying them. We know our sows, we know what we are producing, we do not wish to introduce diseases and it appears to be cheaper to produce our own. However, the other side of the argument is that we are not achieving maximum genetic gain. While we are using high-pointed stock Landrace boars and also have a pure-bred Saddleback boar to keep that side of the line going, I query whether we are really lagging too far behind. Another question is whether we have as much grading potential within the source generation as possible; we may be a couple of years behind compared to buying stock.

MANAGEMENT

Weaning and service

A crucial point with outdoor pigs is that animals are not managed individually but in groups, and at weaning it is essential to reduce variability within a group. Sows are treated according to the average condition of the group; individual service dates are unknown and it is average service date which is taken into account. Some producers try to record service date but I question how successful and worthwhile this is.

It is most important to reduce variability within each group of sows so that animals in similar condition go to the boar at the same time. The means of achieving this depends on the size of unit—the more sows present at weaning, the more choice there is to form sow groups. We form groups of sows in similar condition after weaning, over a fortnight period. We wean once each week within that fortnight which means that we are putting sows into a group running with 3 boars on two occasions. But how many times can sows be put into the group and over what period of time should a group be formed? Over the years we have tried putting sows in 3 and 4 times, and we have decided twice is right for us—we know that the second time sows go into the group is within 2 to 3 days of the first sows having been served, and we also know that theoretically we should feed sows on the basis of a fortnight span after weaning. For each unit it is a question of developing a system that suits individual circumstances. Overall, the more groups that can be formed the less mixing of animals is needed but the higher demand there is on boar power and equipment.

Another contentious issue is the number and size of huts in a paddock. Should there be 7 sows in a hut? Low cost systems with 12 animals per hut tend to result in a high return-to-service rate.

Having sorted the sows and mixed them into groups, it is important to consider feeding. Most producers tend to put sows onto a higher level of feeding after weaning, sometimes for 6 weeks; we only feed at a higher rate for 3 weeks but there is no proven correct regime. Another unanswered question is whether sows should be fed a high density ration at this time. No matter what their stage in the cycle, all of our sows are fed a 15 percent protein ration as we think this is the most economical approach.

Most of our feeding regime from weaning to post-service is based on common sense and a knowledge of what is successful in our situation. We are not sure how long we should feed at a high level, and whether the feeding strategy affects embryo implantation or embryo mortality. Flushing prior to service is difficult when service dates are unknown. On our unit, feeding to regain body condition appears to be effective.

The condition of sows at weaning influences the level of feeding and the length of time that high feed levels are given. Time of year also has an effect. We feed up to about 4 kg of feed per sow per day for 3 weeks after weaning, but other units may feed 5 to 5.5 kg of feed.

As with most aspects of management, there is a great deal of variation between outdoor units in the way sows are fed. On our unit, feed use averages about 1.2 tonnes per sow per year. This is a relatively low figure but it seems that we have managed to feed the right amount at the right time and in the right place. Sow condition overall is the best it has been and we have managed to even out sow condition at weaning.

Boars

We run 3 boars in each paddock, which contains about 15 sows after a fortnight, but other producers run 2 or 4 boars per paddock. We use HY boars, which produce animals with a good conformation, and sows seem to be served effectively. By contrast, other producers consider the Large White is a more effective sire. The general recommendation is that boars should be used until 2 years old, but our animals *work* for 2 years.

Infertility

The biggest breeding problem of outdoor producers is that of summer or autumn infertility. We have had relatively few problems during the last couple of years, which I attribute at least in part to the use of electronic sow feeders. However, 3 to 4 years ago we failed to produce 91 litters over a 22 week period, from October to January. This represented a loss in income from finishing some 860 pigs. We are not sure if the sows put to the boar from May to August failed to get in pig or if the foetuses were reabsorbed.

Farrowing

Our main aim is for the sows to produce as many live piglets as possible. We then try, within a group of sows, to minimise the range in age of the piglets, as it is much simpler to rear and finish piglets 2 to 4 days rather than 10 days apart in age. However, the size of a unit influences the farrowing policy it adopts. Providing that sows are in warm, draught-free huts in the farrowing paddock, the survival of piglets is largely influenced by management.

There are no clear guidelines for the management of farrowing piglets. Increased interest in outdoor pigs in the last 5 years has increased the number of suggested options, without necessarily providing answers. As with most aspects of outdoor pigs, different approaches suit different units. The many, often small, differences in the management of farrowing sows all contribute to the overall success of the unit.

Amongst the questions we ask ourselves are:

* Should huts be insulated?
* Should huts have doors or boards?
* Should huts be moved between farrowings?
* Should the bedding be fully replaced after each farrowing?
* Is it important to keep piglets in the huts until 3 weeks old or is it adequate to keep them in for an extra couple of days after birth?
* Should piglets receive an iron injection?

Producers should record piglet mortality and try to analyse its causes. Important details include whether piglets found dead were born dead or died at or just after birth, and if conditions in the farrowing paddock influenced piglet survival.

The best system for feeding lactating sows is also not clear. Feeding rolls on the ground in a straight line is a simple method, but there is some squabbling amongst the sows and some sows take longer than others to emerge from their huts to eat. We find *ad libitum* feeders very useful and, contrary to some other producers' experiences, feed consumption is not excessive. Since installing *ad libitum* feeders, our sows average about 8 kg feed per day, with little change in cold weather, although feed intake falls slightly in hot conditions. A major advantage of the feeders is that sows are in similar body condition at weaning.

Also contrary to some producers, we have gilts and sows in the same farrowing paddock. Although this often tends to increase variability in the paddock, particularly through competition when feeding on the ground, we find that as we introduce *ad libitum* feeders about 10 days after farrowing, smaller gilts are given an opportunity then to receive their full feed requirement. Some units prefer to put in their *ad libitum* feeders 4 days after farrowing.

PIG HEALTH

On any unit, constant attention must be paid to pig health. We have experienced a number of disease problems over the years, which contributes to our reluctance to purchase gilts. There are different views on whether a policy of preventive medicine should be adopted at all times, or whether diseases should be tackled as they occur. On our unit we vaccinate regularly against SMEDI, Erysipelas and Clostridial scours, but problems such as *E. coli* scours are dealt with when they arise.

Pre-weaning mortality is often the result of poor management, but it can also be due to disease, especially scours. Scouring can occur both in winter, when it is usually attributed to damp conditions, or in summer for other reasons.

LABOUR AND UNIT SIZE

Adequate numbers of good staff are an essential component of successful outdoor pig units. We already have difficulty in finding good calibre staff, and the problem is likely to get worse in the next 3 or 4 years. School leavers and youngsters entering the industry need training and are not fully experienced for 6 to 7 years. A planned labour structure is important, but often disrupted as many people decide to leave when approaching middle age, particularly on an outdoor unit.

Size of unit and level of productivity are closely related to labour requirements. There tends to be greater motivation when one person is managing 200 to 300 sows, and performance levels are often very high. However, the viability of this size of unit is often doubtful, and more sows are certainly needed if a unit is to afford accommodation for finishing pigs rather than selling them as weaners.

A good balance for many larger units seems to be to keep 500 to 700 sows, which requires several staff. Increasing the number of sows offers more opportunity to split the overall pig unit into several smaller units, which increases the promotion prospects which can be offered between and within the different parts of the business. On the whole, I find that an elongated staffing structure is detrimental to animal performance. Correct feeding of sows is crucial to the success of outdoor pig units, and staff must have a close interest in this and pay great attention to the details of management.

I am sure that we have too many sows in one unit and performance would improve if we split the sows into 2 herds. Until now we have expanded as one unit, which has become too large to operate most effectively. One option is to increase sow numbers and to form 2 herds, but a lower cost alternative is for us to split the farrowing paddocks. At the same time, there are units of a similar size to ours which have decided that remaining as one unit is a better option for their circumstances.

CONCLUSIONS

Within outdoor pig production there are many grey areas and a multitude of options are chosen on different units. The success of a unit depends on a combination of appropriate management, the right stock people, cost-effective use of feed and a degree of luck in matters such as the weather and accidents. When problems arise on outdoor units they may be difficult to pin point, not only due to a lack of adequate basic information from research, but also because outdoor units tend to be more influenced than indoor units by external factors such as the climate and land conditions. A great deal of satisfaction can be derived from working with outdoor pigs, but it is not a suitable system for everyone.

DISCUSSION

A possible solution to the problem of lack of finishing pigs due to an outbreak of infertility on a unit would be to take the pigs to a heavier weight, providing that the breeds used are suitable.

The energy as well as the protein level in the feed is important, and the energy component tends to have more influence on price than the protein level. The 15 percent protein ration fed on the Allenford unit has a digestible energy content of 13.5 MJ/kg DM. A more energy dense feed could be given but is not considered worthwhile because, with the *ad libitum* feeders, after farrowing sows can eat to appetite and energy intake should not be a problem. A higher energy feed has been tried in the past but there was no noticeable improvement when using it, and it was wasteful for the dry sows. *Ad libitum* feeders have helped to solve the problem of a considerable loss of body condition after farrowing.

The main reason for expanding from a 200 to 300 sow herd, which is operating very successfully, to a 500 to 700 sow unit, with its accompanying labour problems, is that this number of sows is needed to justify buildings if a producer wishes to progress and produce a bigger pig than a weaner.

Arable farmers often regard outdoor pigs as a disadvantage for their crops, and many would prefer not to have them.

Employing females on outdoor pig units is a contentious issue, particularly as the work is hard and can require considerable strength. At present outdoor pig production is a male-dominated industry, which can lead to practical problems in working conditions for females.

CHAPTER 3

REPRODUCTIVE PROBLEMS AND DISEASES IN OUTDOOR PIGS

A E Wrathall

SUMMARY

In general, pigs kept outdoors are prone to similar types of reproductive problems as those kept indoors. While some of the adverse factors associated with housing systems and intensive production are reduced or removed by keeping pigs outside, other factors increase in importance. Management practices can have a considerable effect on reproductive performance. While some management decisions, such as age at weaning, are common to indoor and outdoor herds, others are more specific to the outdoor situation. Other factors which influence reproductive efficiency include level of stockmanship, genetic characteristics, occurrence of infections, access to poisonous plants, provision of adequate but not excessive levels of nutrients, particularly vitamins and trace elements, and feeding level in relation both to stage of the reproductive cycle and to seasonal weather conditions.

The major difference between indoor and outdoor pigs is the degree to which they are exposed to environmental influences, especially climatic ones. In this paper, therefore, most attention is devoted to results of investigations of seasonal reproductive problems associated with environmental factors.

Modern domestic pigs, unlike their wild ancestor in Europe which is a seasonal breeder, are expected to reproduce efficiently all the year round. Over the past decade, however, there have been many reports of both summer infertility and autumn abortions, which have occurred in various countries. Investigations into these seasonal problems have been carried out mostly with indoor herds but the findings are relevant to pigs kept outdoors. Summer infertility is a problem mainly in countries with hot summers (e.g. southern Europe, U.S.A. and Australia) and researchers have looked particularly at the role of heat stress. Work on the autumn abortion syndrome has been carried out mainly in cooler, higher latitude countries like Canada and Great Britain, and researchers have focussed primarily on the effects of changing light patterns.

Although much remains to be done to understand fully the mechanisms of seasonal reproductive failures in domestic pigs, and also the physiology of normal seasonal breeding in the wild boar, it is already clear that there is no single trigger factor which explains everything. Temperature (heat, cold and diurnal fluctuations), light (day- and night-length and diurnal variation), solar radiation (especially levels causing sunburn), nutrient intakes and other factors (e.g. bedding, body contact, mud wallows) affecting energy balance and homeostasis are just some of the seasonal influences involved. There are, in addition, certain non-seasonal factors (e.g. genetics, toxins and lack of boar-sow social interactions) which, when acting in concert with seasonal ones, probably help to precipitate the summer and autumn reproductive problems. These could be the key to why pigs in some herds are markedly affected whilst others, apparently subject to the same seasonal stresses, are able to reproduce efficiently all-year-round.

INTRODUCTION

Keeping pigs outdoors in a "rich and varied" environment inevitably increases the chance of new and unexpected problems arising. Thus, although outdoor pigs in general have similar types of reproductive problems to animals kept indoors, there tends to be a wider range of causes. While some of the adverse factors associated with housing systems and intensive production are reduced or removed by keeping pigs outside, other factors may increase in importance.

FACTORS AFFECTING REPRODUCTIVE PERFORMANCE

Figure 3.1 illustrates the common categories of factors affecting reproductive problems. Management is a central feature, and several deliberately chosen management policies have greater implications for reproduction than might be originally thought. For example, a decision to wean early, at under 3 weeks, can lead to embryonic losses and cystic ovaries, which tend to result in a reduced subsequent litter size and increased culling of older sows due to infertility.

There are other decisions specifically concerned with keeping pigs outdoors which may also significantly affect reproductive performance. For example, the degree of supervision of service in the paddocks by the stockman, or whether shade, wind shelters and wallows will be provided can have an influence on reproduction.

Apart from deliberate management decisions, other less tangible factors play a role in reproduction, particularly stockmanship. Although much discussed, this is very difficult to define or control; a good stockman is vital, however, for high reproductive performance.

Genetic factors, such as skin colour, are very important with reproduction outdoors. Dark-skinned pigs are less prone to the effects of sunburn, whilst those with a tendency to fatness are more resistant to cold.

Figure 3.1   Common categories of factors affecting reproductive performance

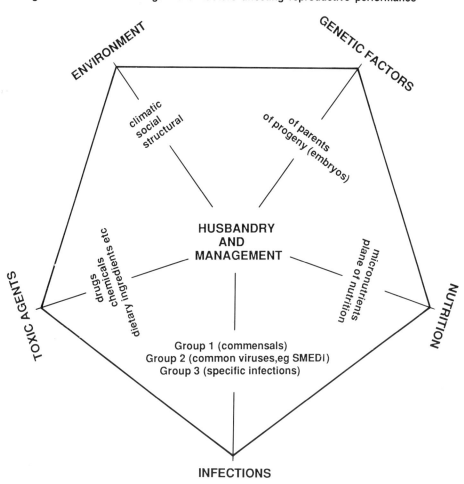

Nutrition also has important effects on reproductive performance, not only with respect to supplying the correct balance of nutrients such as protein and vitamins, but also in relation to the appropriate level of feeding. Feeding level is influential not only at different stages of the reproductive cycle but also at different times of the year. Choosing the correct seasonal feeding level, particularly with regard to energy balance, is particularly important for good reproduction in outdoor pigs, and this is often not adequately appreciated. Some micronutrients, such as vitamins A and D, and inorganic elements like iron and manganese, are less likely to be deficient outdoors because the pigs have access to grass, sunlight and soil. Conversely, exposure to certain toxic elements such as cadmium and selenium, which may occasionally build up in soil, for example after the spreading of sewage sludge, may be more of a threat to the health of outdoor pigs than to those kept indoors.

Infections occur with outdoor pigs as well as with those kept indoors. Organisms such as parvovirus need to be controlled in the same way as for indoor herds. Compared to pigs kept indoors, outdoor animals may be exposed to a much wider range of plants, poisons and parasites. For example, poisonous plants such as hemlock, which causes foetal deformities, and ergotised grasses, which lead to agalactia, may be found in the pasture, while parasite build-up and infections transmissible from wildlife may cause reproductive problems which indoor pigs seldom experience.

A further important factor influencing reproduction is the environment, and a major difference between indoor and outdoor herds is the degree to which animals are exposed to environmental influences. The environment may be considered as having 3 components which are highly relevant to reproduction in both indoor and outdoor herds, namely climate, season and structure. However, climatic and seasonal effects are probably of most significance to outdoor herds.

Modern domestic pigs, unlike their wild ancestor in Europe which is a seasonal breeder, are expected to reproduce efficiently all the year round. Over the past decade, however, there have been many reports from various countries of both summer infertility and autumn abortions. Investigations into these seasonal problems have been carried out mostly with indoor herds, but the findings are relevant to pigs kept outdoors.

SEASONAL EFFECTS ON REPRODUCTION

In June 1986 the EEC sponsored a conference on seasonal effects on reproduction, which was devoted almost entirely to reports of seasonal fertility problems in pigs (Seren and Mattioli, 1987). Delegates from most countries of Northern Europe, such as Denmark, Holland and Ireland, considered that seasonal variations in fertility were minimal, at least in the national pig population, which comprised mainly indoor herds. Results from many thousands of sow records and artificial inseminations showed small differences in days to oestrus after weaning, returns to service and, occasionally, a litter size effect. Overall, there was considered to be little to worry the average indoor pig farmer.

The only aspect of real concern in northern Europe, especially in Britain where the topic has been studied for some time, was the problem of autumn abortions. This is a distinct seasonal problem which, at least in some herds, can be devastating.

In contrast to northern Europe, delegates from some of the southern countries, for example Portugal, Spain, Italy and Greece, reported severe difficulties with returns to service, particularly in mid-summer. In addition, other significant problems in the summer and early autumn were anoestrus after weaning, failure to farrow and abortions, which occurred to a much greater extent than in the UK. Overall, there was a seasonal infertility syndrome covering several different types of reproductive problems.

Although summer infertility as observed in southern Europe does not appear to be a problem in northern Europe, its study is of considerable interest to the UK, as it appears that the underlying pathological mechanisms of the summer infertility and autumn abortion syndromes may essentially be similar.

Summer infertility in southern Europe

Fertility problems in summer manifest themselves in a variety of ways, as previously mentioned, such as delayed puberty, anoestrus after weaning and, most importantly, a reduced farrowing rate. Figure 3.2 shows the seasonal farrowing rate pattern for southern Europe, with farrowing rate plotted against month of *service* (not month of farrowing).

Figure 3.2 Seasonal farrowing rate pattern in southern Europe

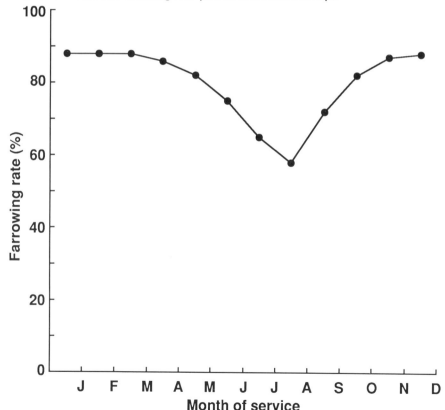

Efficient indoor herds in the UK would expect a farrowing rate of at least 85 to 90 percent. Of the sows which did not farrow (ie. the 'failures') about 7 to 10 percent would be expected to be regular returns, 2 to 3 percent irregular returns, 1 to 2 percent not-in-pig and 0.5 to 1 percent abortions. Figure 3.2 suggests that in the first 4 months and the last 2 months of the year herds in southern Europe perform well, with about 85 to 90 percent of sows farrowing. However, during the summer and early autumn the farrowing rate falls quite dramatically, and for sows which are served in July and August it can be considerably less than 70 percent.

This decreased farrowing rate significantly reduces overall herd efficiency, and its causes have been the subject of much discussion. In the 1960s and 1970s the syndrome was generally attributed to boar infertility due to heat stress, as sperm production in boars which are too hot, even for a few hours, can be severely disrupted. The boar may continue to serve sows, apparently normally, despite having a greatly reduced level of fertility, and nothing will be

Figure 3.3 Average monthly temperatures in Milan and London

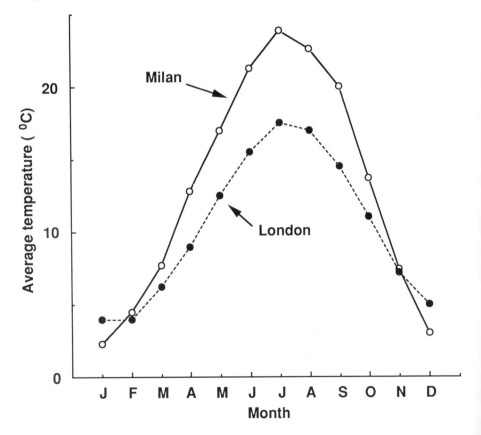

suspected until the sows return to service. Poor fertility may continue for 6 to 8 weeks after an episode of over-heating.

There is little doubt that this is an explanation for some of the reductions in farrowing rates, particularly in early summer when temperatures are rising very fast and fluctuating widely. All boars seem to have great difficulty in adjusting to sudden heatwaves and it appears that, compared to the UK, temperatures in southern Europe can fluctuate more widely. For example, in northern Italy average temperatures in January are 3 to 4°C lower than in the UK, while summer temperatures are higher, and the rate of temperature increase in spring is greater than in the UK (Figure 3.3). This imposes more stress on both boars and sows.

If a low fertilisation rate is the principle factor responsible for summer infertility problems, sows would be expected to show a high percentage of regular returns to service (i.e. with an interval of 3 weeks). From Figure 3.4, which shows the incidence of failure to farrow plotted against month of

Figure 3.4 Incidence of and reasons for sows failing to farrow in southern Europe

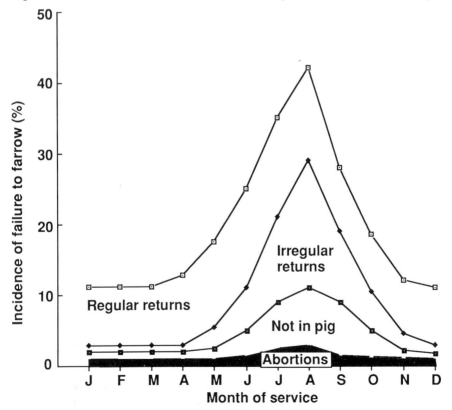

service, it appears that the greatest increase in return rate in the summer is due to irregular returns, with a considerable increase in sows not-in-pig and also a noticeable increase in abortions. As the graph relates to month of service, a considerable time may elapse before animals are recorded as infertile, or as failing to farrow.

The conclusion from Figures 3.2 and 3.4 is that summer infertility is not due only to boar problems. The female must also play a role, as there are increases in sows not-in-pig, abortions and irregular returns, rather than just regular returns.

The question may then be asked as to why some sows have delayed returns to service, some do not return to service and some just never farrow. Do the sows conceive and then lose their embryos, or do they not conceive initially?

An Italian team, led by Professor Seren of the Bologna Veterinary School, used the oestrone pregnancy test to address these questions (Mattioli, Prandi, Camporesi, Simoni and Seren, 1987). This very accurate test, which can detect living embryos at about 25 days, revealed that about 50 percent of sows in the 'irregular return' and 'not-in-pig' groups actually had live embryos at 25 days, which then seemed to disappear. At the same time, an almost equal proportion of sows were diagnosed as not pregnant at 25 days. Thus it appears that several factors are involved in the summer infertility syndrome and much more research is needed to clarify the exact mechanisms involved.

Summer infertility in the UK

In the UK a few herds, especially outdoor herds, experience a form of summer infertility. Unfortunately, there has been little research on summer infertility in outdoor herds and few data are available. Consequently information tends to be extrapolated from southern Europe.

Summer infertility is not confined to outdoor animals and incidences in indoor herds have been reported. For example, 3 herds in Cornwall which used kennel and yard systems had what appeared to be a typical summer infertility problem, that is a large number of irregular returns occurred during the summer months (Hancock, 1988). Figure 3.5 shows the percentage of irregular returns in one of the herds over a period of 3 years. There appeared to be a correlation between irregular returns and hours of sunlight, and by protecting sows both from sunlight and from excessive heat, through modifying the housing, there was a marked reduction in the number of returns to service. A similar situation was observed for the other two herds, and changes in housing also had marked effects with these.

Another example, with different clinical manifestations, was reported in 1982 by a veterinary surgeon in Thetford, Norfolk (John Wilkinson, personal

Figure 3.5   Incidence of irregular returns in a pig herd in Cornwall, recorded over 3 years

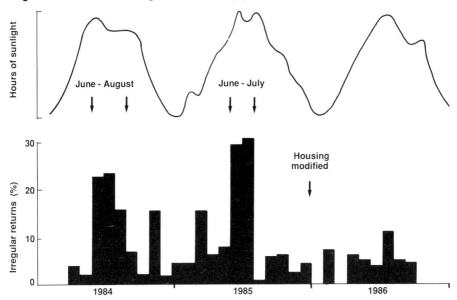

Source: Hancock (1988).

communication at Pig Veterinary Society meeting). This was a very unusual summer abortion problem in outdoor pigs, with abortions occurring in large numbers over a very short period. The production system involved sows being mated indoors and kept inside until they had passed their 3-week return date, before being turned out into paddocks. This system worked well during the winter and spring, until the time of turn out coincided with hot sunny weather in June. At this point many of the pigs were very badly sunburnt, and almost immediately afterwards a large proportion of the sows aborted. A number of theories were proposed, but there is very little other scientific evidence to link sunburn and abortion either in pigs or other animals.

Autumn abortion syndrome

On a typical pig unit about 0.5 to 1 percent of pregnancies might be expected to abort (Table 3.1). Although abortions can occur at any time, there appears to be a definite increase in autumn in northern countries such as the UK and Canada. In the UK the syndrome typically varies in severity from year to year; in a bad year some herds may experience up to 10 percent abortions, while with other herds abortions are rarely seen. Most abortions reported to the Central Veterinary Laboratory (CVL) seem to occur in indoor herds, but this may be an artefact due to greater ease of detection in indoor rather than outdoor herds. Although the overall level is not high, abortions are a very tangible aspect of a seasonal reproductive inefficiency which is noticed by producers.

Table 3.1 Expected values for farrowing rate and incidence of failure to conceive or farrow in "efficient" pig herds

	Incidence (%)
Farrowing rate	85–90
Regular returns	7–10
Irregular returns	2–3
Not-in-pig	1–2
Abortions	0.5–1

Autumn abortions show few distinctive clinical features. Sows can be affected at any stage of pregnancy, and they seldom show any signs of ill health, not even loss of appetite either before or after the abortion. In the aborted litters, the foetuses tend to be equal in size and their size is appropriate for their gestational age. Apart from occasional subcutaneous haemolysis, there are no obvious pathological lesions. These rather negative manifestations all suggest the sudden failure of pregnancy control through maternal failure, and foetal disease is not implicated.

Figure 3.6 VIDA II pig fetopathy diagnoses

(Classification of over 11 000 submissions, 1975 to 1984)

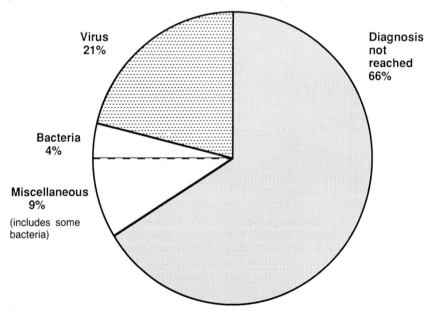

Virus 21%

Diagnosis not reached 66%

Bacteria 4%

Miscellaneous 9%

(includes some bacteria)

Source: MAFF Central Veterinary Laboratory.

Many theories have been proposed over a number of years, ranging from infection, climatic stress, feed toxins, and under-feeding to loss of the hair coat. However, to date there has been no conclusive evidence to support any theory and a single factor cannot be implicated as the sole cause.

In Britain all abortion submissions which are sent for diagnosis to MAFF Veterinary Investigation Centres are recorded on the Veterinary Investigation Diagnosis Analysis (VIDA) system, which allows analysis of the records. There have been several thousands of foetopathy (aborted or mummified foetus) submissions to VIDA, and as a first step foetuses were classified into 4 broad groups by cause of the abortion: bacterial, viral, miscellaneous and no diagnosis (Figure 3.6).

Figure 3.7   VIDA II pig fetopathy: average monthly number of submissions for which no diagnosis was reached, 1975 to 1984

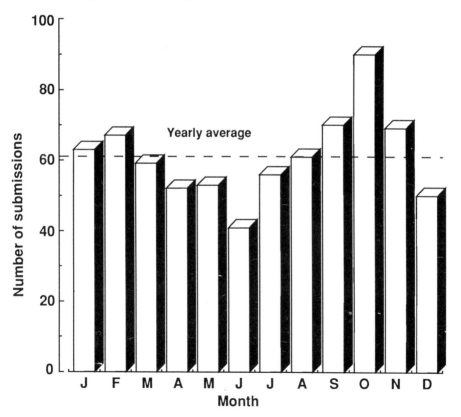

Source: MAFF Central Veterinary Laboratory.

Most of the cases of foetopathy which were diagnosed were of viral origin. Of these, most would have been due to parvo-virus infection, and were thus not true abortions. The category of 'no diagnosis' is the one where most of the real abortions are, in fact, classified and over a 12 year period there was a definite seasonal pattern of this category of abortions, with a peak in October (Figure 3.7). This appears to represent the autumn abortion syndrome, suggesting that infection is probably not involved in most cases.

The next step taken by MAFF in the search for reasons for the autumn abortion syndrome was to analyse the records for patterns in herds which had problems with autumn abortions. Figure 3.8 shows the pattern of abortions in a herd with a high abortion rate. In 3 successive years, there were peaks of abortions clustered towards the end of the summer, around the month of September. Overall, the CVL had precise dates for 270 abortions and, with the help of the Meteorological Office, correlations were sought between these dates and climatological changes such as temperature, atmospheric pressure, day length and sunlight. The only correlation found was with day length. Figure 3.9 shows the incidence of abortions in 1978, together with the *rate of change* in day length, which is greatest towards the end of September. The data showed that abortion dates were most strongly correlated with the rate of change of day length, not day length *per se*, suggesting that rapidly declining day length may have a causative role in the autumn abortion syndrome.

Figure 3.8   Seasonal occurrence of abortion in a pig herd with a high abortion rate

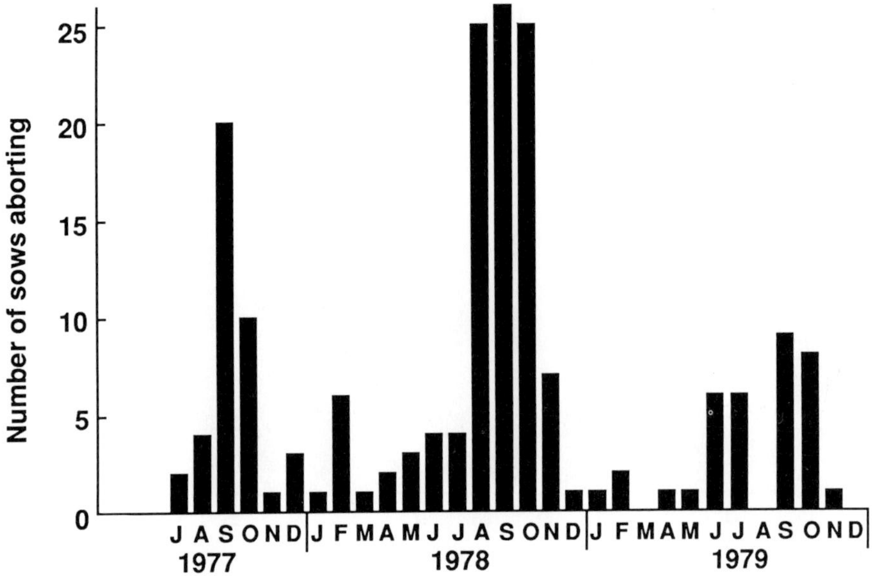

Source: MAFF Central Veterinary Laboratory.

Figure 3.9 Occurrence of abortions in 1978 in relation to rate of change in daylength

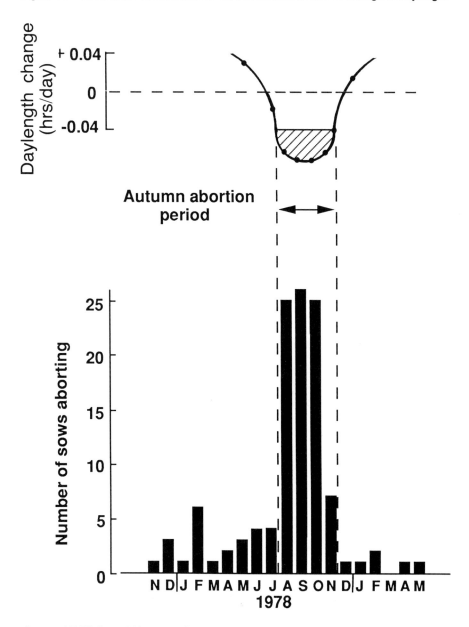

Source: MAFF Central Veterinary Laboratory.

Work in France on the wild pig has shown that its reproductive activities halt about July, which is when day length begins to decline. Then, depending on food supply and other factors such as social interactions, reproductive activity recommences between October and December. It seems reasonable to consider that in modern domestic pigs autumn abortions and other seasonal problems may be a reflection of an ancestrial tendency to seasonal breeding. If this is true, seasonal variations in hormone levels of domestic adult breeding pigs would be expected.

Figure 3.10   Location of farms in the autumn abortion project

Of the various endocrine mechanisms of abortions, especially those of the maternal failure type, I believe the one most likely to be involved with the autumn abortion syndrome is reduced luteotropic support from the pituitary glands. The *corpora lutea* of pregnancy in the pig seem to be dependent upon luteinising hormones, and in pregnant sows progesterone levels are a direct reflection of luteinising hormone (LH) secretion.

The next stage in the CVL study was to carry out a blood sampling survey on 7 farms scattered around the country, to investigate if there was a seasonal decline in pregnancy hormones. At each of the 7 farms (denoted A to G in Figure 3.10) samples of blood were taken from 20 pregnant sows in September, December, March and June, to measure progesterone levels. Table 3.2 gives hormone levels in autumn, winter, spring and summer, from which it can be seen that progesterone concentrations in autumn were significantly lower than at other times of the year on most farms.

The CVL also studied the seasonal endocrinological effects in an eighth herd (Herd H; Figure 3.10). In this herd of 250 sows, over a period of 12 months most sows became pregnant twice, and each was blood sampled twice per pregnancy—first at 25 to 30 days and later at 70 to 90 days. Figure 3.11

Table 3.2   Seasonal comparisons of mean (± SE) serum progesterone concentrations (ng/ml) in pregnant sows in seven herds[1,2]

	Season of sampling			
Herd	Autumn	Winter	Spring	Summer
A	9.70 ± 1.88 (15)	23.10 ± 1.07*** (25)	15.25 ± 1.25* (23)	16.29 ± 1.28** (38)
B	22.40 ± 1.98 (21)	30.83 ± 1.50** (29)	17.46 ± 1.89 (17)	25.53 ± 1.32 (76)
C	16.52 ± 1.42 (27)	31.23 ± 1.80*** (27)	17.88 ± 0.91 (30)	23.15 ± 1.52** (29)
D	13.54 ± 0.67 (20)	25.38 ± 1.22*** (21)	17.99 ± 1.22** (20)	20.31 ± 1.38*** (23)
E	16.43 ± 1.13 (24)	22.14 ± 1.88* (22)	23.86 ± 1.04*** (24)	21.25 ± 2.20 (18)
F	5.77 ± 1.75 (13)	17.86 ± 2.79** (11)	18.24 ± 1.59*** (8)	16.83 ± 2.39** (11)
G	13.29 ± 0.97 (17)	22.85 ± 1.51*** (20)	27.51 ± 1.54*** (19)	25.06 ± 2.62*** (16)
All herds	14.80 ± 0.67 (137)	25.73 ± 0.70*** (155)	19.75 ± 0.59*** (141)	22.12 ± 0.70*** (211)

[1] Number of samples are given in brackets.

[2] Values significantly different from the autumn mean are: *   p <0.05
                                                              ** p <0.01
                                                              *** p <0.001

36

shows the mean results of almost 1000 blood samples taken during the year. For both early and late pregnancy samples there was a definite seasonal trend in progesterone levels, with a decrease between about June and October and an increase from November. It was also clear that *corpora lutea* which had formed in the summer and autumn were very competent to produce more progesterone, but their low progesterone production seemed to indicate a seasonal reduction in luteotropic stimulation. Thus, an important factor appears to be associated with lack of stimulation from the pituitary gland, more of which would have led to greater progesterone production by the *corpora lutea*.

Figure 3.11   **Seasonal variations in serum progesterone levels in pregnant sows in a 260-sow herd**

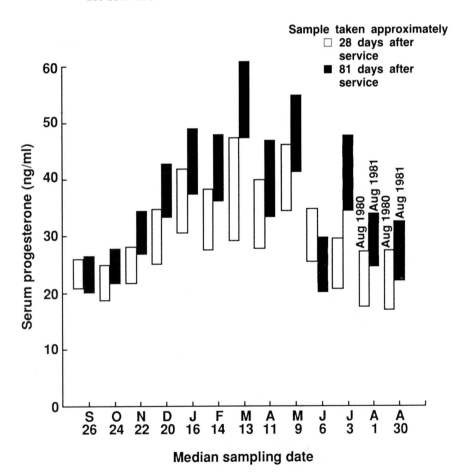

Low progesterone secretion does not fully explain the autumn abortion syndrome because the great majority of sows do not actually abort in the autumn. Other adverse factors are presumably needed to tip the balance between the continuation of pregnancy and abortion.

CONCLUSIONS

Figure 3.12 summarises the current position. The top part of the figure, above the dotted line, shows what is probably a normal seasonal effect for all sows, namely a rapid decline in day length, reduced luteotrophic support to

Figure 3.12 Seasonal effects on reproduction in sows

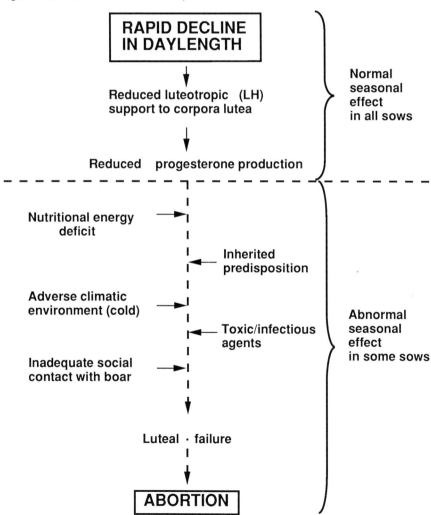

38

the *corpora lutea* and a reduced progesterone level. These changes probably occur in all sows, and it is only when other adverse factors are superimposed that the events shown in the remainder of the figure occur, leading to luteal failure and abortion.

Observations on reproductive failure at the CVL and in Canada suggest that nutritional energy deficit is very important, as is an adverse climatic environment. These two factors are interrelated in terms of energy balance, and their effects may be modified by other factors such as bedding and body contact. In addition, certain non-seasonal factors such as genetics, toxins and lack of boar-sow social interactions (which can boost progesterone levels), acting in concert with seasonal ones, probably help to precipitate summer and autumn reproductive problems. These could be the key to why pigs in some herds are markedly affected whilst others, apparently subject to the same seasonal stresses, are able to reproduce efficiently all-year-round.

REFERENCES

Hancock, R.D. (1988) Seasonal infertility in sows in Cornwall. *The Veterinary Record*, **123**, 413-416.

Mattioli, M., Prandi, A., Camporesi, A., Simoni, A. and Seren, E. (1987) Investigations on swine summer infertility in Italy. In: Seren, E. and Matioli, M. (Eds) *Definition of the Summer Infertility Problem in the Pig.* A seminar in the Community programme for the coordination of agricultural research, held at the Istituto di Fisiologia Veterinaria, Bologna, 19 and 20 June 1986. Commission of the European Communities, Agriculture, EUR 10653 EN, 33-43.

Seren, E. and Mattioli, M. (Eds) (1987) *Definition of the Summer Infertility Problem in the Pig.* A seminar in the Community programme for the coordination of agricultural research, held at the Istituto di Fisiologia Veterinaria, Bologna, 19 and 20 June 1986. Commission of the European Communities, Agriculture, EUR 10653 EN.

DISCUSSION

Although very little work has been carried out on outdoor herds, there is evidence that seasonal infertility, in particular summer infertility, is a greater problem in outdoor herds compared to those kept indoors.

Once the causes of seasonal reproductive problems are understood, the next step is to determine what can be done to control the problems. Administering hormones has only been tried with sows kept indoors, and mainly in relation to the autumn abortion syndrome. Pig producers in southern Europe are very interested in the, as yet, unanswered question of whether sows can be given progesterone and HCG (human chorionic gonodotrophin) injections to reduce summer infertility and abortions.

CHAPTER 4

ASPECTS OF REPRODUCTIVE ENDOCRINOLOGY IN OUTDOOR PIGS

J M Bassett

SUMMARY

The endocrine mechanisms regulating reproduction are the same whether pigs are kept outdoors or inside. However, environmental influences have a much more direct impact on outdoor pigs and may be less readily ameliorated. The principal environmental factors influencing reproduction in outdoor pigs can be classified as photoperiodic, thermoregulatory, psychological/sociological/behavioural (stressors) or nutritional, and changes or disturbances in any of these factors may influence the neuroendocrine systems regulating reproductive activity. The principal endocrine regulators of reproductive activity in pigs are considered to be well-established. However, recognition that opioid-like peptides are involved in regulating gonadotrophin release and that the gonads also produce peptides such as inhibin and activin, which have important feed-back effects on pituitary gonadotrophin production, is leading to reassessment of the way gonadotrophin release is regulated during the reproductive cycle. Further, interest in the role of local growth factors within the ovary is shedding new light on the mechanisms by which nutritional alterations can influence follicular development. Greater understanding of these mechanisms has increased interest in the interactions between the system directly regulating reproduction and the signals generated by the aforementioned environmental factors.

Because of the economic importance of summer infertility, particular interest has been shown in the mechanisms by which seasonal factors influence reproduction. Although domestic pigs are far less distinctly seasonal breeders than wild pigs or other species such as sheep, clear evidence indicates that important changes in reproductive activity, related to changing photoperiod, occur. In most species examined so far, changes are mediated through plasma melatonin and may be accompanied by variations in plasma prolactin concentration. By contrast, although alterations in photoperiod may improve fertility in pigs, seasonal rhythms of melatonin are poorly defined and it is not clear whether they play any role in regulating seasonal changes in pig reproduction. Further studies are needed. Extremes of environmental temperature also have a seasonal incidence, and may amplify and confound photoperiodic effects.

The impact of nutrition, considered in Chapters 5 to 7, on reproduction should not be forgotten.

INTRODUCTION

Seasonal problems of reproduction in outdoor pigs are encountered relatively frequently worldwide, as outlined in Chapter 3. However, while seasonally-related problems in reproduction may have an endocrine basis, little endocrine research specifically related to the outdoor pig has been carried out. Attempts to assess the physiological and hormonal mechanisms which may be involved therefore rely heavily on published observations on other species, or on pigs kept in controlled environments. While the hormonal mechanisms regulating reproduction are the same whether pigs are kept indoors or outdoors, environmental influences have a more direct impact on outdoor animals, are more variable and may be less easily ameliorated.

The principal environmental factors influencing reproduction may be classified as:

* Photoperiodic;
* Thermoregulatory;
* Psychological (sociological or behavioural stressors);
* Nutritional.

Changes or disturbance in any of these factors may influence the neuroendocrine systems regulating reproductive activity. Many interactions among these various factors undoubtedly occur when reproductive failure is precipitated. However, to begin to appreciate the significance of observations on the effects of these environmental factors on the reproductive process, it is necessary to provide an outline, at least, of the way that reproduction in the pig is regulated hormonally, so that mechanisms through which these factors may act can be pinpointed.

In this paper attention will be mainly concentrated on the sow, although reference will also be made to the boar. It should be recognised, however, that environmental influences have just as great an impact on the boar as the sow and their consequences for herd fertility may be greater because of the relative numbers of the two sexes in the herd.

HORMONES AND THE CONTROL OF REPRODUCTION IN ADULT PIGS

Once pigs reach an appropriate body size, puberty and subsequent regular cyclical reproductive activity are controlled through variations in the production of the gonadotrophins, follicle-stimulating hormone (FSH) and luteinising hormone (LH), produced within the anterior pituitary gland situated at the base of the brain. These hormones travel through the blood to the ovaries and testes, where they regulate production of ova and sperm. Increased functional activity in the gonads results in the production of a number of hormones characteristic of the sex, which, in their turn, bring about changes elsewhere in the body.

In the female the best known of these gonadal steroid hormones are oestrogen and progesterone, while in the male the best known is the androgen testosterone. Alterations in the production of these hormones regulate the oestrus cycle, prepare the uterus for pregnancy, maintain pregnancy and regulate the growth and initiation of functional activity in the mammary glands. In the male, testosterone is concerned with supporting spermatogenesis and with regulating the expression and maintenance of the secondary sexual characteristics essential to successful reproduction. These hormones also have very important 'feed-back' effects within the brain, where their concentrations regulate production of the neuropeptide releasing factor which controls production of the gonadotrophins by the anterior pituitary gland.

Neuroendocrine control of reproductive activity

The neuroendocrine system controlling reproduction in sows and boars is illustrated schematically in Figure 4.1. The most important region of the brain involved in regulating the function of the anterior pituitary gland is the hypothalamus, which is situated immediately above it in the base of the brain. While the function of the posterior pituitary gland, which produces the hormones oxytocin and vasopressin, is regulated through nerve fibres passing directly to it from the hypothalamus, control of the production of the gonadotrophins (FSH and LH) as well as that of the other anterior pituitary hormones—prolactin (PRL), growth hormone (GH), thyrotrophin and adreno-corticotrophin (ACTH)—is achieved by secretion of regulatory factors (some stimulatory and others inhibitory) into blood vessels of the portal system running from the hypothalamus to the anterior pituitary. There are a large number of such regulatory factors released into the portal system, from neuroendocrine cells located elsewhere in the hypothalamus.

For the regulation of reproduction, the most important of these releasing factors is gonadotrophin-releasing hormone (GnRH), which is synthesised within the cell bodies of neuroendocrine neurones, located mainly in the medial basal region of the hypothalamus. This releasing factor is secreted from axonal processes which terminate adjacent to the hypophysial portal vessels. GnRH then travels to the anterior pituitary, where it controls the secretion of FSH and LH.

It is known that release of GnRH from the hypothalamus is intermittent, or pulsatile, and that both the frequency and amplitude of the bursts of secretory activity have very important roles in regulating secretion from the pituitary gonadotroph cells. The steroid hormones from the gonads have important roles in regulating these patterns of release and it is the variation in their production which is primarily responsible for the changes in GnRH, and thus pituitary gonadotrophin release, throughout the cycle. The details are complex and a more comprehensive text, such as that by Karsch (1984), should be consulted for a fuller account.

Figure 4.1 The major regulatory systems involved in the neuroendocrine control of reproduction and their relationships

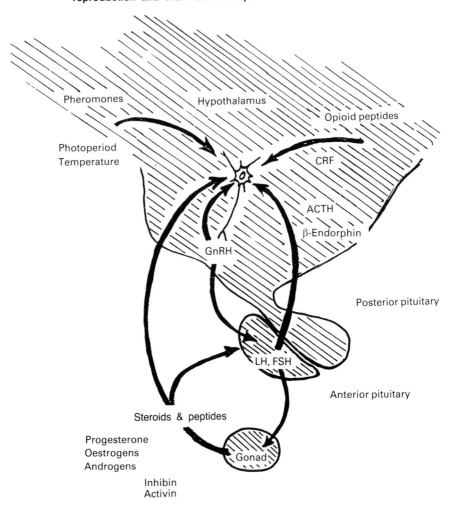

Secretion of gonadotrophins (FSH and LH) from the anterior pituitary gland is controlled by gonadotrophin-releasing hormone (GnRH) from the hypothalamus region of the brain. GnRH release is controlled by environmental factors via neural, neuroendocrine (eg. opiates, endorphins) and endocrine (eg. corticotrophin-releasing factor (CRF), adrenocorticotrophic hormone (ACTH)) pathways. Hormone release from both hypothalamus and pituitary is subject to feed-back regulation by steroid and peptide hormones released from the gonads as a result of gonadotrophin action.

Source: Modified from Karsch (1984).

In addition to the regulatory influence of the hypothalamus, there are a considerable number of other factors referred to later, including the gonadal steroids, which separately influence secretion of LH and FSH. However, it is the release of GnRH by the hypothalamus which is crucial to the control of the reproductive cycle in both sows and boars. As control of GnRH release is a neuroendocrine mechanism, there are inputs to the GnRH-secreting neurones from other parts of the hypothalamus and also from other regions of the brain. For example, it is well-known that pheromones such as androstenone (boar odour) have an important influence on reproductive activity through the olfactory system.

In this connection, recent studies on mice by Schwanzel-Fukuda and Pfaff (1989) have shown that GnRH-producing neurones originate in the nasal region and then, during early embryonic life, migrate to the hypothalamus through the olfactory tract of the developing brain. However, they leave behind some processes and cells. The GnRH neurones relocated in the hypothalamus clearly retain their connection with the olfactory system and its sensory elements in the nose, providing an important higher level for control of reproductive activity in sows.

Important inputs to the GnRH neurones in the hypothalamus also come from other hypothalamic centres involved with the regulation of body temperature and the generation of daily and seasonal rhythms in response to the light-dark cycle and variations in daylength.

Control of the oestrus cycle by the anterior pituitary

The production of FSH and LH and their relation to alterations in the production of oestradiol and progesterone throughout the ovarian cycle are shown schematically in Figure 4.2. The most important feature is the increase in blood levels and secretion of LH and FSH associated with the manifestation of oestrous behaviour, and the maturation and ovulation of follicles at this time. Although in this generalised picture the size of the peaks in FSH and LH are shown as equal, the increase in FSH in sows is actually far less than that in LH, and the duration of the period when progesterone levels are low is 4 to 5 rather than 2 to 3 days, as shown.

During the remainder of the cycle changes in FSH and LH are more independent of one another, the changes being determined largely by changes in the negative feed-back exerted on the hypothalamus and anterior pituitary by varying levels of the gonadal steroids, progesterone and oestrogen. Progesterone in increasing amounts decreases the frequency of GnRH and LH pulses, while increasing oestradiol levels decrease their amplitude. The changes in steroid levels result from the growth and maintenance of corpora lutea, which form after ovulation from the ruptured follicles and which produce progesterone, or from successive waves of

44

Figure 4.2   A generalised schematic representation of follicular and endocrine changes from luteal regression to resumption of luteal function in domestic livestock

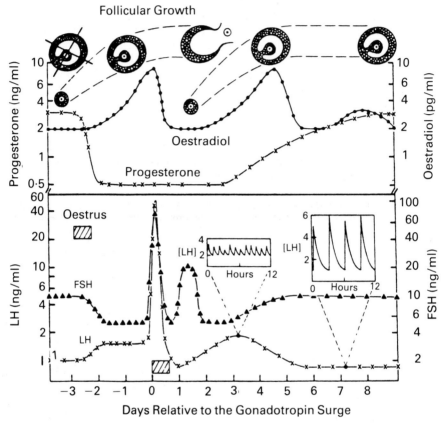

Growth and regression of oestrogen-producing follicles are shown in the top panel. No attempt has been made to represent the number of follicles during each cycle in the sow.

Source: Hansel and Convey (1983).

follicles which grow and produce oestrogen, but then regress, until signals from the uterus, if a fertile mating has not occurred, promote regression of the corpora lutea. The withdrawal of the high level of progesterone then permits increased secretion of LH, a transition to positive feedback effects of oestrogen which increase the amplitude of LH secretory pulses and promote development of new ovulatory follicles. As the follicles grow, FSH secretion begins to decline under an inhibitory feed-back influence of inhibin, a peptide which they produce.

Although the actual levels of LH and FSH are of significance during the oestrus cycle, it appears that their pattern of change may be of greater significance. Recent data on hormone levels in sows selected on the basis of differences in ovulation rate (Kelly, Socha and Zimmerman, 1988) show that sows which produced a much larger number of eggs had higher levels of FSH, both before and in the first few days after oestrus, but although the differences were statistically significant, they were relatively small. A marked feature, however, was that the pattern of hormone production around ovulation was very constant for all animals, suggesting that other features of the system may have been more important determinants of the differences in ovulation rate.

Figure 4.3 illustrates schematically some of the events occurring in the ovary during the oestrus cycle and, for comparison, similar events occurring in the testis during the spermatogenic cycle.

Figure 4.3 **Diagrammatic representation of the interrelationships between systemic hormones (FSH and LH) and local growth factors (IGF-1), TGF-β, EGF, inhibin, activin) in regulating steroidogenesis (E2 oestradiol, T testosterone, A androstendione, ABP androgen-binding protein) and follicular growth in the ovary (a) or spermatogenesis in the seminiferous tubules (b)**

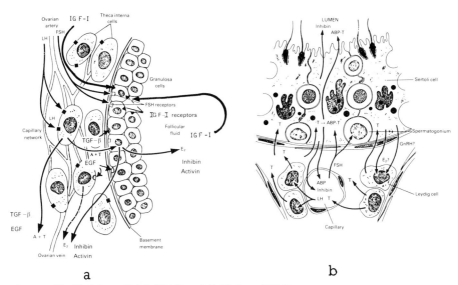

a   b

Source: Modified from Baird (1984) and DeKretser (1984).

Local control of hormone action and production within the gonads

The ovary contains a large number of inactive follicles. During each oestrus cycle a small proportion of these primordial follicles begin to grow, although not all of them actually mature to produce ovulatory follicles. There appear to be mechanisms within the ovary which provide local control of these events.

Increasing LH levels result in a rapid increase in ovarian blood flow, but the concentration of LH in blood reaching the ovary is possibly not a critical factor in itself. Rather, the distribution of blood within the ovarian tissue may be a more important determinant of the way LH regulates follicular development. Figure 4.3a shows the complex network of blood capillaries which develops around a follicle as it begins to grow and which provides the follicle and the thecal cells around it with nutrients and hormones required for their development. The ability of the ovary to control blood flow to individual follicles may be a critical factor in determining the rate at which this capillary network develops and whether a particular follicle will grow and mature. The possibility that this occurs is well illustrated by recent observations of Brown and Driancourt (1989), which showed that follicular blood flow in a highly fecund breed of sheep is far greater than that in a breed with lower fecundity, despite little difference between the breeds in blood flow to the underlying stromal tissue (Table 4.1).

Table 4.1   Blood flow in the ovaries and ovarian follicles of high fecundity Romanov and low fecundity Prealpes-du-Sud sheep

	Sheep breed	
	Romanov	Prealpes-du-Sud
Stromal blood flow	32.8 ± 9.2	40.8 ± 4.4
(ml/min/100 gram)		
Arteriovenous shunts (%)	0.3 ± 0	60.2 ± 16.3
Follicle blood flow		
(ml/min $\times$ 10^4/mm^3) (total flow)		
1 to 3 mm diameter	522 (116)	269 (77)
3 to 5 mm diameter	385 (313)	197 (174)
5 to 7 mm diameter	203 (311)	261 (517)
Number of follicles per ewe	3.8	1.6

Source: Brown and Driancourt (1989)

Opening of arteriovenous shunts within the ovary evidently plays a major role in diverting blood away from developing follicles, but the mechanisms involved remain to be determined. Similar measurements have not so far been reported on pigs.

The vascular system also delivers other hormones besides FSH and LH to the ovary, but possible roles for them have not been considered by reproductive physiologists until recently. It is now clear, however, that insulin-like growth factor-1 (IGF-1), which appears to mediate many of the effects of GH at the tissue level, and a number of other growth factors have important effects on the multiplication and hormone responsiveness of cells around the follicle and within the avascular compartment of the follicle itself. Experiments indicate that these growth factors, besides stimulating cell multiplication, alter the number of receptors for FSH and LH on theca and granulosa cells and induce increased synthesis of enzymes limiting their ability to synthesise oestrogens. They, therefore, have important effects on the ability of developing follicles to secrete oestrogens.

There is also considerable evidence that the granulosa cells themselves produce IGF-1 in response to FSH and other growth factors produced by the surrounding thecal cells. These latter cells also produce several growth factors, adding to the complexity of the situation. Alterations in the production of local growth factors in the ovary may be an important way by which the sow's nutritional state influences her reproductive efficiency.

In addition to producing growth factors concerned with regulating the multiplication and differentiation of cells which provide the specialised functions to regulate production of steroid hormones characteristic of the ovary (Figure 4.3a), granulosa cells also produce peptides which have feed-back effects on the pituitary. The best known of these is inhibin, which plays a role within the granulosa of the follicle, but which is also secreted into the venous blood leaving the ovary. Inhibin from the ovary feeds back onto the anterior pituitary to inhibit selectively the secretion of FSH. Another structurally-related peptide, activin, also produced by granulosa cells, is thought to promote FSH secretion from the pituitary. The significance of these ovarian peptides for control of the reproductive cycle is still being investigated.

In the testis, regulation of spermatogenesis within the seminiferous tubules (Figure 4.3b) by FSH and, indirectly, by LH via its effect on the production of testosterone in the interstitial cells adjacent to the tubules has many similarities to regulation within the ovary. Many other hormones besides the pituitary gonadotrophins are involved and the system of control is a multi-layered one involving production of local growth factors and other regulators. Inhibin is also produced to influence events within the tubules and to provide a very important negative feed back onto the anterior pituitary via the vascular drainage of the organ.

Local regulation by a variety of growth factors, only a few of which have been mentioned, clearly plays an important role within the gonads in modulating effects of pituitary gonadotrophins on the proliferation and differentiation of ovarian follicles and spermatogonia. Growth factors involved in the gonads have been the subject of recent detailed reviews (Carson, Zhang, Hutchinson, Herrington and Findlay, 1989; Bellve and Zheng, 1989).

Pregnancy and parturition

During a normal oestrus cycle ovulation will be followed by the luteinization of the ruptured follicles, resulting in the increased synthesis and secretion of progesterone, as shown in Figure 4.1. About two weeks later, if pregnancy does not ensue, the corpora lutea will begin to regress and a new cycle will start.

In the non-pregnant sow a prostaglandin (PGF2alpha), produced by the uterine endometrium and secreted into the the uterine vein, is transferred to the ovarian artery by a local countercurrent mechanism and brings about this regression of the corpora lutea. However, if fertilization of ova occurs and pregnancy is established, oestrogen secreted from the expanding blasto-cysts alters the way in which uterine tissue processes the PGF produced by the endometrium. Instead of the PGF2alpha entering the venous blood and being carried to the ovary, it is diverted elsewhere, so levels reaching the ovary are much reduced and luteolysis prevented. The production of oestrogen by the trophoblast and its effects on PGF are illustrated in Figure 4.4. For a fuller description of these mechanisms Bazer and First (1983) should be consulted.

Because of this mechanism, and in contrast to many other species, oestrogens appear to contribute to the maintenance of the corpus luteum during early pregnancy in sows. However, LH secretion from the pituitary of the sow remains the principal luteotrophic agent throughout pregnancy. Nevertheless, since the maintenance of pregnancy is totally dependent on continued production of progesterone by corpora lutea throughout pregnancy in sows, inadequate uterine production of oestrogen may be the factor most likely to lead to failure of the pregnancy during its early stages. Indeed, recent studies (King and Rajamahendran, 1988) suggest that low amplitude infrequent pulses of prostaglandins, produced by the endometrium, may prevent corpora lutea from achieving their maximum progesterone secretory potential in all pregnant pigs.

Parturition in the sow follows luteolysis resulting from increased production of PGF2alpha by the uterus, consequent on maturation of the hypothalamo-pituitary adrenal axis in the fetal piglets; increased production of cortisol by their adrenals leads to greatly accelerated uterine production of the prostaglandins and to the entire complex of events associated with withdrawal of the progesterone block to myometrial contractility. This subject has been reviewed comprehensively by First, Lohse and Nara (1982) and by Liggins (1982).

Lactation and weaning

After delivery, the inhibitory effect of progesterone from the corpora lutea on gonadotrophin stimulation of ovarian follicular development is replaced by reflex neural inhibition of GnRH release and pituitary responsiveness to GnRH initiated by suckling. While the intensity of suckling remains high, this

Figure 4.4   Diagrammatic representation of pregnancy prolonging the life of the corpus luteum in the sow

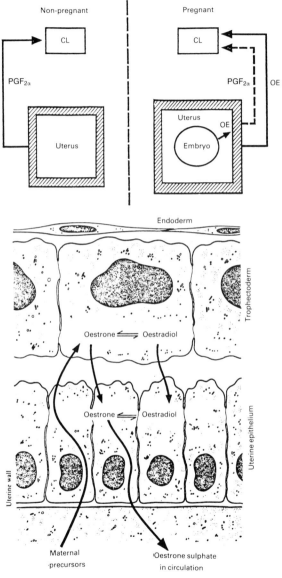

Expansion of the trophoblast within the uterine lumen leads to production of oestrogen (OE), which inhibits release of prostaglandins (PGF) into the uterine vein, preventing luteolysis. Oestrogens themselves are further transformed within the uterine wall.

Source: Heap and Flint (1984).

reflex inhibition of gonadotrophin release remains sufficient to prevent the resumption of normal ovarian cyclical activity in the sows, just as in many other species. However, as suckling frequency decreases, the suppression of hypothalamic GnRH and pituitary LH release becomes less severe. Experimental investigations have also shown that the response of LH secretion to stimulation by GnRH increases steadily as lactation progresses. Nevertheless, few sows in outdoor herds would be expected to resume oestrus activity before weaning at 3 or 4 weeks.

In detailed investigations of the effects of lactation and weaning on the secretion of the gonadotrophins, Shaw and Foxcroft (1985) have shown that infrequent, large amplitude, episodic pulse releases of LH do occur during lactation (Figure 4.5). In some animals (eg. sow a) pulses of LH release occur at 3 hourly intervals, while in others (eg. sow b) there may be only one pulse in a 12 hour period. This is despite very high levels of prolactin, which are frequently associated with suppression of GnRH and LH release.

Studies by others have shown, however, that the complete suppression of prolactin secretion has little influence on the responsiveness of LH release to GnRH stimulation. It is quite clear that the pattern of LH release changes dramatically after weaning and that the time taken to return to oestrus is related to the pattern generated during this period (Figure 4.5). Shaw and Foxcroft (1985) reported that the length of the weaning to oestrus interval was inversely correlated with the mean LH concentrations during the period before weaning, although there was no similar correlation with FSH levels, either before or after weaning. However, the ovulation rate at the first post-weaning oestrus was positively associated with the level of FSH in the immediate post-weaning period.

The mechanisms mediating the inhibitory influence of lactation (suckling) on hypothalamic GnRH release remain obscure. Increased production of endogenous opioid peptides within the hypothalamus or other areas of the brain has been implicated as a possible mediator of the suppressive effect of suckling on GnRH and LH release, since it has been shown in a number of species that administration of the opioid-antagonist, naloxone, increases plasma LH. Recent studies by Armstrong, Kraeling and Britt (1988) show that administration of the opioid agonist, morphine, to lactating sows results in reduced LH levels and in suppression of the marked increase in pulsatile LH release which follows removal of their litters for 8 hours (transient weaning). Furthermore, additional experiments by the same investigators on sows weaned after an 8 to 9 week lactation showed that thrice-daily injections of morphine for 5 days after weaning increased the interval to oestrus by approximately 5 days, providing the first evidence that endogenous opiate peptides could be used to inhibit uniformly, and thereby synchronize, the return of sows to oestrus.

Figure 4.5 Plasma FSH, LH and prolactin in sows weaned at 21 days post partum, with weaning to oestrus intervals of a) 4 days; b) 6 days and c) > 11 days

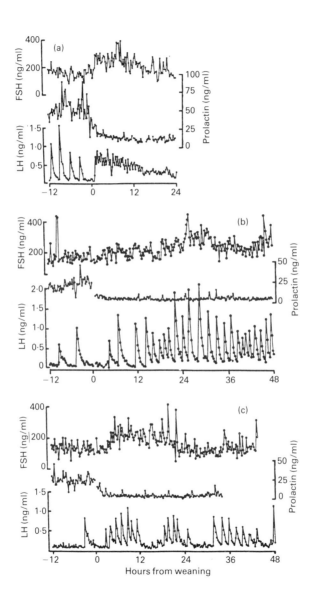

Source: Shaw and Foxcroft (1985).

The experiments also provide further circumstantial support for the theory that endogenous opiate peptides, probably produced within the central nervous system, are involved in modulating the influence of suckling on the suppression of LH release during lactation.

There is therefore considerable evidence that central neural regulatory mechanisms play an important role in determining when sows will return to normal oestrus activity after the cessation of lactation.

ENVIRONMENTAL FACTORS INFLUENCING REPRODUCTION

Seasonal alterations in photoperiod

The very clear evidence that seasonal factors can have very important effects on reproduction in domestic pigs kept indoors and outdoors is described in Chapter 3, and has been the subject of extensive earlier reviews (Claus and Weiler, 1985; Hennessy, 1987; Seren and Mattioli, 1987). However, while evidence from the industry tends to indicate that environmental temperature, and in particular high summer temperatures, may be the principal cause of adverse effects on reproduction in pigs, studies by Mauget and associates (Mauget and Boissin, 1987; Ravault, Martinat-Botte, Mauget, Martinat et al, 1982) on European wild pigs which have a clearly defined seasonal pattern of breeding indicate that varying daylength, rather than temperature, is the principal environmental determinant of the seasonal rhythm in their reproductive activity.

In this respect, the wild pig is similar to sheep which have a marked seasonal period of anoestrus related to long-day photoperiods. However, it is not at all clear whether the loss of a clear seasonal period of anoestrus with domestication in the pig has resulted from similar genetic changes to those which must have occurred in sheep, where some breeds such as the Merino have little, if any, true seasonal anoestrus period, while others such as the hill breeds of Britain have long periods of seasonal anoestrus, comparable to those of primitive types such as the Soay or Mouflon. Differences in seasonal infertility on reproductive activity amongst modern pig breeds remain poorly defined, but may exist.

Despite the absence of a clearly defined seasonal anoestrus period in domestic pigs, the results of Armstrong, Britt and Cox, 1986; Claus and Weiler (1987) and Wheeler (1987) provide clear evidence for the existence in domestic pigs of rhythms in reproduction and reproductive hormone levels, closely related to the seasonal rhythm of daylength. Claus and Weiler (1987) reported clear seasonally-related rhythms in measures of libido in AI boars, in addition to plasma and semen concentrations of testosterone, while they, and Armstrong et al (1986) observed marked increases in the length of the weaning to oestrus interval in sows during summer. Further, as shown in

Figure 4.6   Seasonal variations in plasma testosterone of AI boars subject to a natural photoperiod or to a light reversal pattern (left panels), or in the weaning to oestrus interval in sows under natural daylight or subjected to a light-decrease programme from May to September (right panel)

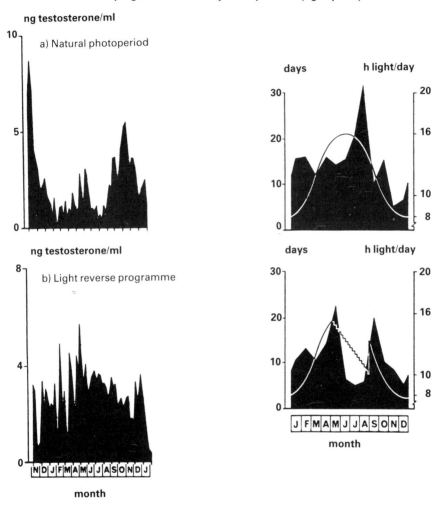

Source: Claus and Weiler (1987).

Figure 4.6, Claus and Weiler have demonstrated that light reversal programmes, or the introduction of a programme of decreasing daylength from May onwards, can alter the normal changes seen during summer, clearly demonstrating their relationship to photoperiod.

In sheep a very marked increase in plasma prolactin level is associated with the period of seasonal anoestrus, and similar changes are seen in plasma prolactin concentrations of wild pigs (Ravault *et al*, 1982). No similar clearly defined rhythm of prolactin has been observed in domestic pigs. On the other hand, Wrathall (1987) observed a well defined seasonal rhythm of plasma progesterone in sows, both early and late in gestation, and Armstrong *et al* (1986) observed significant differences between sows slaughtered in spring or in summer in the GnRH content of their hypothalami and LH concentration in their pituitary glands.

A further marked difference between pigs and other photoperiodically-sensitive mammalian species also appears to exist in the way in which photoperiodic responses are mediated. In the sheep and other species which have been investigated, it is well established that photoperiodic changes are

Figure 4.7 **Comparison of diurnal melatonin patterns in sows subjected to different lighting regimes (a) with those in ewes (b) adapting to a short-day photoperiod (8 hour light) after prolonged exposure to long days (16 hour light)**

a   b

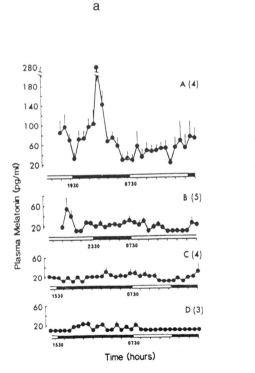
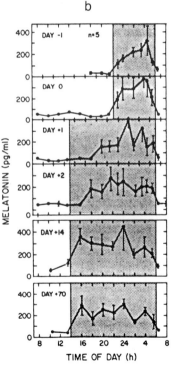

Sources: (a) McConnell and Ellendorff (1987).
         (b) Bittman *et al* (1983).

monitored in the body through their effect on the diurnal rhythm of melatonin secretion from the pineal gland.

The complex pathways regulating transfer of information from the eyes to the brain and thence to the pineal to control melatonin synthesis are well established, and have been reviewed by Arendt (1986). At present, however, despite a number of investigations, there is still no incontrovertible evidence that a diurnal rhythm in plasma melatonin concentration plays any part in modulating the effects of changing daily photoperiod in the pig. McConnell and Ellendorff (1987) have reported the existence of a diurnal rhythm in plasma melatonin in pigs kept under a 12:12 hour light:dark cycle, but they were unable to show that the duration of the period of melatonin release was related to the period of darkness when the daily photoperiod was altered. The contrast between their observations and those of Bittman, Dempsey and Kassch (1983) on sheep subjected to a change in the photoperiod to which they were exposed (Figure 4.7) illustrates clearly the difference between observations on the two species.

Unfortunately, information on plasma melatonin levels in pigs is still very limited, so it remains difficult to assess whether the difference is due to methodological problems with pig plasma samples, or whether the pig, amongst mammals investigated so far, is unique in not relying on a melatonin rhythm for coordination of its diurnal and seasonal rhythms. Further investigations are clearly essential if the mechanisms by which photoperiod influences pig reproduction are to be elucidated.

Seasonal alterations in environmental temperature

Apart from obvious consequences such as sunburn, keeping pigs in hot climates or outdoors during the summer in more temperate regions may lead to physiological adaptations which will markedly alter the distribution of blood flow from the heart to the tissues. For example, blood flow to the skin will be increased in an effort to increase heat loss from the body surface, and in consequence flow to internal organs may be reduced. However, although there is little evidence that blood flow to the ovaries or the uterus is greatly altered, there is considerable evidence that severe heat stress does cause marked alterations in the reproductive endocrinology of both sows and boars (Wetteman and Bazer, 1985).

Australian studies (Paterson and Pett, 1987) indicate that high environmental temperatures immediately before or within 15 days of mating are a major cause of seasonal infertility and suggest that changes in the endocrine milieu of the sow at this time are responsible. In this context it is noteworthy that recent evidence suggests that the uterine changes preventing release of prostaglandins into the uterine vein may fail if uterine tissues become hyperthermic.

There is substantial evidence for impaired spermatogenesis when boars are exposed to high temperatures for prolonged periods of time (Wettemann and Bazer, 1985; Cameron, 1987). Measurements of normal intratesticular temperatures show that these are 2.5°C below rectal temperatures. The maintenance of this lower temperature is of considerable importance, as studies on many species show that cryptorchidism or local heating of the testis to normal body temperature leads to severe abnormalities in spermatogenesis, with pachytene spermatocytes being particularly susceptible (Setchell, 1982). Other studies indicate that testosterone secretion may also be impaired. Because of the long duration of the testicular cycle, weeks may elapse before adverse effects of a period of hot weather show as reduced boar fertility.

PSYCHOLOGICAL FACTORS

Stressful factors in the environment are also recognised as influencing the reproductive behaviour of pigs through effects on the hypothalamo-pituitary-adrenal axis. However, while Hennessy (1987) has reviewed this subject, the role of environmental stresses leading to activation of the pituitary adrenal axis in promoting reproductive failure remains somewhat controversial.

CONCLUSIONS

The extended outline of endocrine changes associated with regulation of reproduction in pigs has indicated the complexity of the processes involved. However, attempts to understand how environmental changes result in failure in the process of reproduction ultimately depend on utilizing information on many aspects of pig physiology, and it seems unlikely that successful control of the seasonal infertility problems of outdoor pigs will result from investigation of their endocrinology alone without consideration of many other aspects of their physiology and husbandry.

REFERENCES

Arendt, J. (1986) Role of the pineal gland and melatonin in seasonal reproductive function in mammals. *Oxford Reviews of Reproductive Biology*, **8**, 266-320.

Armstrong, J.D., Britt, J.H. and Cox, N.M. (1986) Seasonal differences in function of the hypothalamic-hypophysial-ovarian axis in weaned primiparous sows. *Journal of Reproduction and Fertility*, **78**, 11-20.

Armstrong, J.D., Kraeling, R.R. and Britt, J.H. (1988) Morphine suppresses luteinizing hormone concentrations in transiently weaned sows and delays onset of estrus after weaning. *Journal of Animal Science, **66**, 2216-2223.

Baird, D.T. (1984) The Ovary. In: Austin, C.R. and Short, R.V. (Eds) *Reproduction in Mammals: 3 Hormonal Control of Reproduction.* Cambridge University Press, Cambridge, 91-114.

Bazer, F.W. and First, N.L. (1983) Pregnancy and parturition. *Journal of Animal Science*, **57**, Supplement 2, 425-460.

Bellve, A.R. and Zheng, W. (1989) Growth factors as autocrine and paracrine modulators of male gonadal functions. *Journal of Reproduction and Fertility*, **85**, 771-793.

Bittman, E.L., Dempsey, J. and Karsch, F.J. (1983) Pineal melalonin secretion drives the reproductive response to daylength in the ewe. *Endocrinology*, **113**, 2276-2283.

Brown, B.W. and Driancourt, M.A. (1989) Blood flow in the ovaries and ovarian follicles of Romanov and Prealpe-du-Sud ewes. *Journal of Reproduction and Fertility*, **85**, 317-323.

Cameron, R.D.A. (1987) Effects of heat stress on boar fertility with particular reference to the role of the boar in seasonal infertility. In: *Manipulating Pig Production.* Australasian Pig Science Association, Werribee, Australia, 60-66.

Carson, R.S., Zhang, Z., Hutchinson, L.A., Herrington A.C. and Findlay, J.K. (1989) Growth factors in ovarian function. *Journal of Reproduction and Fertility*, **85**, 735-746.

Claus, R. and Weiler, V. (1985) Influence of light and photoperiodicity on pig prolificacy. *Journal of Reproduction and Fertility*, Supplement 33, 185-197.

Claus, R. and Weiler, V. (1987) Seasonal variations of fertility in the pig and its explanation through hormonal profiles. In: Seren, E. and Mattioli, M. (Eds) *Definition of the Summer Infertility Problem in the Pig*, Commission of the European Communities, Luxembourg 1987, 127-139.

DeKretser, D.M. (1984) The testis. In: Austin, C.R. and Short, R.V. (Eds) *Reproduction in Mammals: 3 Hormonal Control of Reproduction.* Cambridge University Press, Cambridge, 76-90.

First, N.L., Lohse, J.K. and Nara, B.S. (1982) The endocrine control of parturition. In: Cole, D.J.A. and Foxcroft, G.R. (Eds) *Control of Pig Reproduction.* Butterworths, London, 311-341.

58

Hansel, W. and Convey, E.M. (1983) Physiology of the estrous cycle. *Journal of Animal Science*, **57**, Supplement 2, 404-424.

Heap, R.B. and Flint, A.P.F. (1984) Pregnancy. In: *Reproduction in Mammals: 3 Hormonal Control of Reproduction*. Cambridge University Press, Cambridge, 153-194.

Hennessy, D.P. (Ed) (1987) Symposium: Seasonal infertility in the pig. In: *Manipulating Pig Production*. Australasian Pig Science Association, Werribee, Australia, 40-73.

Karsch, F.J. (1984) The hypothalamus and anterior pituitary gland. In: Austin, C.R. and Short, R.V. (Eds) *Reproduction in Mammals: 3 Hormonal Control of Reproduction*. Cambridge University Press, Cambridge, 1-20.

Kelly, C.R., Socha, T.E. and Zimmerman, D.R. (1988) Characterisation of gonadotropic and ovarian steroids during the periovulatary period in high ovulating select and control line gilts. *Journal of Animal Science*, **66**, 1462-1474.

King, G.J. and Rajamahendran, R. (1988) Comparison of plasma progesterone profiles in cyclic, pregnant, pseudopregnant and hysterectomised pigs between 8 and 27 days after oestrus. *Journal of Endocrinology*, **119**, 111-116.

Liggins, G.C. (1982) The fetus and birth. In: Austin, C.R. and Short, R.V. (Eds) *Reproduction in Mammals: 2 Embryonic and Fetal Development*. Cambridge University Press, Cambridge, 114-141.

McConnell, S.J. and Ellendorff, F. (1987) Absence of nocturnal plasma melatonin surge under long and short artificial photoperiods in the domestic sow. *Journal of Pineal Research*, **4**, 201-210.

Mauget, R. and Boissin, J. (1987) Seasonal changes in testis weight and testosterone concentration in the European wild boar (*Sus scrofa* L.). *Animal Reproduction Science*, **13**, 67-74.

Paterson, A.M. and Pett, D.H. (1987) The role of high ambient temperature in seasonal infertility in the sow. In: *Manipulating Pig Production*. Australasian Pig Science Association, Werribee, Australia, 48-52.

Ravault, J.P., Martinat-Botte, F., Mauget, R., Martinat, N., Locatelli, A. and Bariteau, F. (1982) Influence of the duration of daylight on prolactin secretion in the pig: hourly rhythm in ovariectomized females, monthly variation in domestic (male and female) and wild strains during the year. *Biology of Reproduction*, **27**, 1084-1089.

Schwanzel-Fukuda, M. and Pfaff, D.W. (1989) Origin of luteinising hormone-releasing hormone neurons. *Nature*, **338**, 161-164.

Seren, E. and Mattioli, M. (Eds) (1987) *Definition of the Summer Infertility Problem in the Pig*. Commission of the European Communities, Luxembourg.

Setchell, B.P. (1982) Spermatogenesis and spermatozoa. In: Austin, C.R. and Short, R.V. (Eds) *Reproduction in Mammals: 1 Germ Cells and Fertilisation*. Cambridge University Press, Cambridge, 66-101.

Shaw, H.J. and Foxcroft, G.R. (1985) Relationships between LH, FSH and prolactin secretion and reproductive activity in the weaned sow. *Journal of Reproduction and Fertility*, **75**, 17-28.

Wetteman, R.P. and Bazer, F.W. (1985) Influence of environmental temperature on prolificacy of pigs. *Journal of Reproduction and Fertility*, Supplement 33, 199-208.

Wheeler, G.E. (1987) Reproductive problems in outdoor pigs. *Pig Veterinary Society Proceedings*, **18**, 41-61.

Wrathall, A.E. (1987) Investigations into the autumn abortion syndrome in British pig herds. In: Seren, E. and Mattioli, M. (Eds) *Definition of the Summer Infertility Problem in the Pig*. Commission of the European Communities, Luxembourg, 45-62.

CHAPTER 5

NUTRITION OF OUTDOOR PIGS

W H Close

SUMMARY

Outdoor sows can be as productive as those kept indoors, but their nutritional requirements are different. Conventionally, nutrient requirements can be calculated as the summation of the requirements for maintenance, some maternal gain, foetal development and milk production. The higher feed requirements of outdoor sows result from their higher maintenance energy requirements, which are associated with a greater level of activity and the more variable environmental conditions to which they are exposed. Although sows graze and eat some straw provided for bedding, it is unlikely that these roughages provide sufficient nutrients to reduce significantly their requirements for compound feed.

Outdoor sows consume 10 to 15 percent more feed than indoor sows and are normally given diets of lower energy content, so that higher feeding levels are needed to provide similar energy intakes. Higher feeding levels are needed in winter than in summer, and an additional 0.6 MJ digestible energy, or about 50g feed per day, should be provided for each 1°C fall in temperature below the critical temperature. In summer, on the other hand, feed intake of the lactating sow may be reduced if the environmental temperature is too high and this may influence piglet growth, by reducing milk production, and subsequent reproductive performance, by extending the weaning to re-mating interval and by reducing litter size. A 1°C increase in temperature has been shown to reduce feed intake in lactating sows by at least 200g per day.

Gilts require a feed intake of approximately 30 MJ digestible energy per day in order to achieve a 30 to 40 kg increase in maternal body weight during their first parity. During the second parity their energy requirements increase to 33 MJ digestible energy per day. In lactation, the objective should be to feed ad libitum, *and sows need to consume at least 90 MJ digestible energy per day to maintain their body reserves. Following lactation, feed intake should remain high until the last sow in the group has been served. This will assist in restoring any loss of body condition during lactation, and will hopefully increase ovulation rate and litter size in the subsequent parity.*

INTRODUCTION

As with other aspects of outdoor pig production, published information on the nutrition of animals kept outdoors is very sparse. From data collated by the

various recording systems, however, it seems clear that the feed requirements of outdoor pigs are considerably greater than those of indoor pigs.
Table 5.1 presents data collected by the Meat and Livestock Commission
(MLC) over the last 6 years, which clearly show that outdoor herds use about
10 percent more feed than indoor ones. The data provided by the Cambridge
Pig Management Scheme recording system, discussed in Chapter 1, also
suggest much greater feed requirements for outdoor herds—some 15
percent higher compared with indoor herds.

Table 5.1 Feed usage by indoor and outdoor herds

Year	Feed usage (tonne/sow/year)		% difference between herds
	Indoor	Outdoor	
1983	1.16	1.20	3
1984	1.14	1.23	8
1985	1.14	1.25	10
1986	1.14	1.26	11
1987	1.16	1.26	9
1988	1.15	1.29	12

Source: MLC Yearbooks (1983-88).

Wheeler (1987) reviewed reproductive problems in outdoor pigs and
highlighted the importance of correct nutrition for optimum sow productivity.
Management and husbandry were similar in the various herds surveyed but
there was a considerable difference in feed intake. Figure 5.1 illustrates the
effect this had on performance, as measured by the number of pigs born per
sow per year, which rose from 17 to 20 with an increase in feed intake.

In order to understand how nutrition influences sow productivity, it is first
necessary to review the nutritional requirements and responses of the
animals at the various stages of reproduction.

PRINCIPLES OF NUTRITION

Feeding pre-breeding

The importance of rearing gilts appropriately so that they are in the correct
condition to have a long reproductive life must be emphasised, and feeding is
an important aspect of this. Table 5.2 shows the effect of protein
concentration in the diet on daily liveweight gain and age at puberty. Low
protein diets are believed to lead to more fat and less lean in the body
compared with higher protein diets, and this may have the effect of
increasing age at puberty. While there is very little available information on
the feeding of modern gilts, Table 5.2 indicates that body composition and
age at puberty can be manipulated by nutritional means.

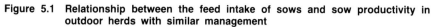

Figure 5.1 Relationship between the feed intake of sows and sow productivity in outdoor herds with similar management

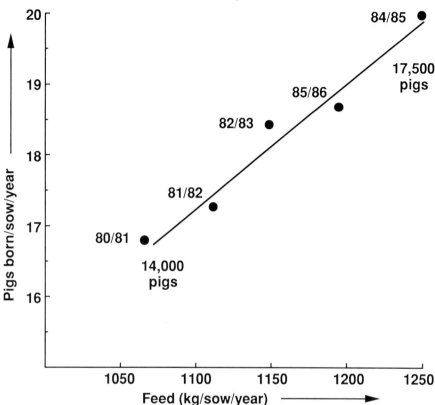

Source: Wheeler (1987).

Table 5.2 Effect of protein adequacy on age at puberty

	Diet 1	Diet 2
Protein content of feed (g/kg)	100	140
Liveweight gain (kg/day)	0.54	0.63
Age at puberty (days)	179	160

Source: Cunningham, Naber, Zimmerman and Peo (1974).

Feeding at mating

The effect of feeding level during the oestrus period on ovulation rate is shown in Table 5.3. Litter size is dependent upon the number of eggs which are shed during ovulation, and upon embryonic mortality. The data in Table 5.3 show that by manipulating nutrition an increase in ovulation rate can be achieved. This is the 'flushing' effect, which is very specific, as the optimum number of eggs shed seems to be dependent upon the time at which feed levels are increased during the oestrus cycle; a general recommendation is two weeks before mating.

Table 5.3 The influence of level of energy intake on ovulation rate in the pig

Energy intake (MJ ME/day)	Ovulation rate	Source
22.5	12.3	Anderson and Melampy (1972)
41.7	14.1	
15.6	11.8	Hartog and van Kempen (1980)
24.9	13.2	

Data in Table 5.3 are relatively old and most modern gilts would probably be fed close to *ad libitum*, thus energy intakes would be expected to be high during the critical period, obviating the specific need to flush.

The next stage is to ensure that as many fertilised eggs as possible survive. Feeding can have a big effect on this, as shown in Table 5.4. There was a considerable improvement in embryo survival when feed intake was reduced during the first 10 days of pregnancy, that is up to the time of implantation. After this there was only a small benefit when feed intake was reduced. However, it is important not to over-restrict following mating as this may result in a lower conception rate.

Table 5.4 Feed intake and reproductive performance in early pregnancy

Days 0-10		Days 10-30	
Feed intake (kg/day)	Embryo survival (%)	Feed intake (kg/day)	Embryo survival (%)
4.1	66	4.1	67
2.5	72	2.5	72
1.25	78	1.25	72

Source: Dutt and Chaney (1968).

Feeding during pregnancy

Nutrient requirements change during pregnancy, and feeding during this period can have a marked effect on reproductive performance. Figure 5.2

demonstrates how nutrient requirements change during pregnancy and the consequences of providing too low a level of feeding to meet the animal's needs. Sows were fed a constant 1.8 kg feed per day throughout pregnancy, in a thermoneutral environment.

Figure 5.2 Partition of energy during pregnancy for sows on a low feeding level (20 MJ ME or 21 MJ DE/day)

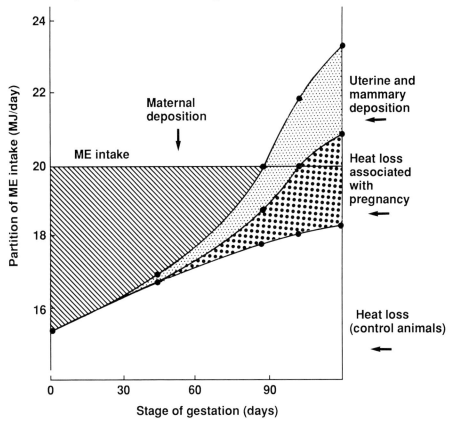

Source: Close and Cole (1986).

The needs of the animal can be partitioned into several components. Firstly, there is a component of energy associated with the amount of heat lost from the body, and as the animal gets bigger the amount of heat lost from the body increases. A second component is the energy required for deposition within the uterus and mammary gland. The difference between heat production plus this value and energy intake in the feed is deposited in the body of the sow. This deposition is high in early pregnancy, but there comes a point when the level of feeding is too low to supply all the nutrients needed during pregnancy. This means that, in order to meet the animal's needs and for sound growth and development of the foetus, mobilisation of body tissue, particularly fat, occurs.

In this experiment, in a thermoneutral environment of 20°C, the sow used her own body reserves to maintain the growth and development of the foetus from day 90 of pregnancy. In colder conditions, however, such as those frequently encountered by the outdoor sow in winter, the point at which maternal energy deposition ceased and body fat was used to meet the need to sustain foetal growth would have occurred at an earlier stage in pregnancy. The application of low feeding levels to sows in cold conditions gave rise to the phenomenon of the 'thin sow' syndrome, which occurred widely in the 1960s and which resulted in a malnourished animal unable to breed.

Figure 5.2 refers to the partition of nutrients in animals on a fixed level of feeding throughout pregnancy. Because of the change in the animals' needs, various strategies have been worked out to feed pregnant sows more efficiently, according to their metabolic requirements. For example, feed may be given at a lower level during the first period of pregnancy and then

Figure 5.3 Partition of energy as fat in the pregnant sow fed low and high levels of energy during gestation

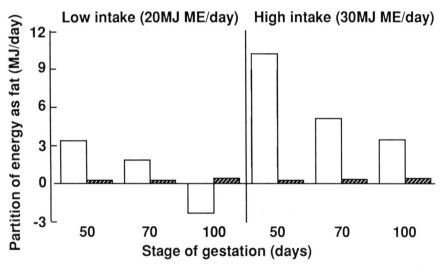

Total fat is partitioned into fat retained in the maternal body (☐) and the conceptus tissue (▨)

Source: Close, Noblet and Heavens (1985).

increased during the later stages of pregnancy, to try to compensate for the rapid increase in heat production and foetal and mammary growth.

Similarly, supplementation of diets in late pregnancy is practised in order to try to overcome the mobilisation of fat and other tissue which occurs in late pregnancy.

Figure 5.3 shows the partition (storage or mobilisation) of fat during gestation when animals were fed at two levels of energy intake, either 20 or 30 MJ metabolisable energy (ME) per day. At both the high and low energy intakes, fat deposition decreased during pregnancy.

At the high feeding level the animals were always able to retain fat in both the maternal body of the sow and the conceptus tissue. However, at the low feeding level, the sows mobilised body fat from their body in late gestation, even though there was a positive gain in the conceptus tissue. It is because of this probable loss of body reserves that fat has been fed in late gestation as a compensatory mechanism.

The extent to which protein intake influences reproductive performance is illustrated in Table 5.5, which gives the results from a large coordinated trial carried out 10 years ago. The data show that feeding similar amounts of energy but varying the protein content of the ration influenced the body weight gain of the sow. With a low protein diet there was less increase in body weight than with a high protein diet because the animal deposited more fat and less lean tissue in its body. There was also a small, although insignificant, increase in the number of piglets born, but little effect on the birthweight of the piglets. Thus the protein content of the diet affected the performance of the sow more than the piglets.

Table 5.5 Effect of level of protein during pregnancy on sow performance

| | Dietary protein level (%) | | | |
	9	11	13	15
Body weight change (kg)	+12.8	+18.7	+21.0	+21.7
Litter size (piglets)	10.1	10.7	10.9	11.0
Piglet birth weight (kg)	1.35	1.34	1.30	1.29

Source: Greenhalgh, Elsley, Grubb, Lightfoot et al (1977).

Lactation

The sow needs to eat as much as possible during lactation. The information in Figure 5.4 is derived from a large number of trials, and shows the change in body weight of the sow and piglets at different digestible energy (DE)

68

intakes. It clearly demonstrates that, in order to maintain sow liveweight during lactation, a very high energy intake is required—over 90 MJ digestible energy per day—which would relate to a daily feed intake of 7 to 8 kg feed. It is questionable whether many sows have the capacity to consume this amount of feed consistently throughout lactation. In practice, the aim should be to ensure that lactating sows eat as much as possible in order to minimise body weight loss. Thus feeding should be *ad libitum*.

Figure 5.4 **Body weight change of sows and piglets in relation to digestible energy (DE) intake during lactation, taken from various sources**

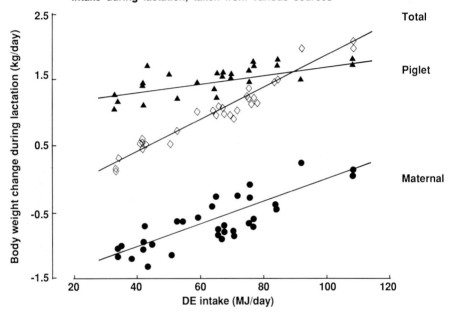

Figure 5.5 further demonstrates the effects of sow nutrition during lactation. When feeding levels were high (70 MJ DE/day) or low (35 MJ DE/day), sows suckling 6 or 12 piglets tried to maintain milk yield by mobilising body reserves. On a high but not a low feeding level, sows with 6 piglets could maintain body tissue, while with 12 piglets sows on both feeding levels had to mobilise substantial amounts of body reserves to meet metabolic needs.

The changes in body composition of the lactating sows in this experiment are illustrated in Figure 5.6. With a low level of feeding and a large number of piglets, sows lost up to 1.5 kg body weight per day over a 21-day lactation, of which half was lean tissue and half fat. It was previously thought that sows mobilised only fat, but it is now known that substantial amounts of lean tissue are also mobilised.

Figure 5.5   The partition of metabolisable energy (ME) by sows during early lactation (days 5 to 13)

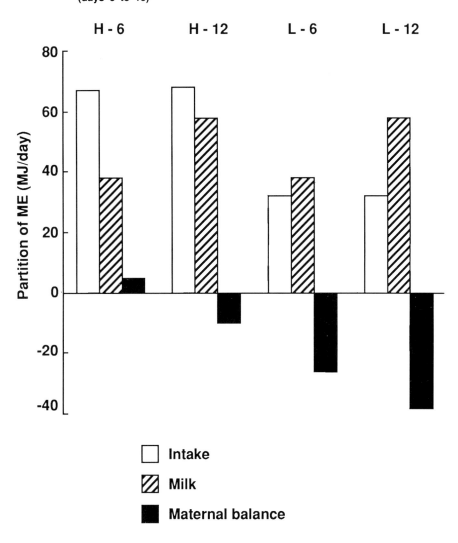

Notes: H = High energy intake (70 MJ DE/day)
         L = Low energy intake (35 MJ DE/day)
         6   refers to animals with 6 piglets
         12   refers to animals with 12 piglets

Source: Mullan and Close (1989).

Figure 5.6 Calculated changes in the body composition of sows during lactation (days 5 to 21)

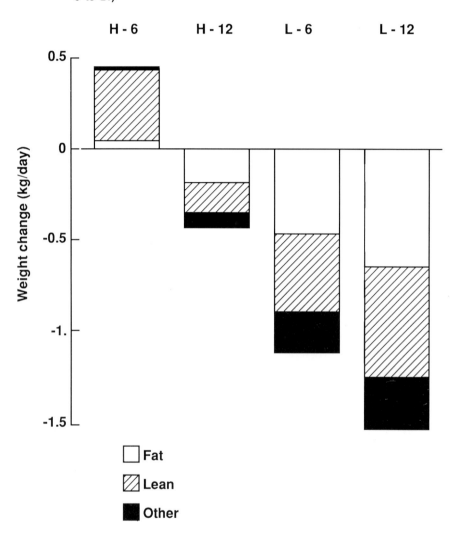

Notes: H = High energy intake (70 MJ DE/day)
L = Low energy intake (35 MJ DE/day)
6 refers to animals with 6 piglets
12 refers to animals with 12 piglets

Source: Mullan and Close (1989).

Results from coordinated sow trials carried out by Greenhalgh, Baird, Grubb, Done et al (1980), given in Table 5.6, show how protein intake had a greater influence on the body weight change of the sow than it did on either piglet performance in terms of numbers weaned or piglet weaning weight. It should be noted that the piglets were mostly weaned at 6 weeks of age, not 3 to 4 weeks as is currently practised, and in these experiments they had access to creep feed from 3 weeks of age. As with nutrition in pregnancy, these data again show that protein intake has more effect on the weight change of the sow during lactation than on the performance of the piglets.

Table 5.6 Effect of level of protein during lactation on sow performance

| | Dietary protein level (%) | | | |
	13	15	17	19
Body weight change (kg)	−6.1	−4.1	−2.6	−1.1
Number of piglets weaned	9.2	9.5	8.8	9.3
Piglet weaning weight at 6 weeks of age (kg)	10.7	10.9	11.3	11.4

Source: Greenhalgh, Baird, Grubb, Done et al (1980).

Temperature also has a considerable effect on the performance of sows and their litters, as shown in Table 5.7. Sows kept indoors were exposed to temperatures from 18 to 30°C. At very high temperatures, for example 30°C, there was a 50 percent depression in feed intake compared with the colder environment. Thus, one of the effects of animals being kept under very hot conditions is a large loss of appetite.

As a result, there were large changes in the body weight of the sow, with animals losing up to 24 kg over a 4-week lactation. Although the piglets had access to creep feed, they did not compensate for a probable decrease in milk supply by eating more creep feed. Thus, the weaning weight of the litter decreased by about 10 kg when animals were kept at 30°C rather than 18°C.

Table 5.7 The effect of temperature on the performance of sows and their litters

| | Temperature (°C) | | |
	18	25	30
Feed intake (kg/day)	6.5	6.1	4.2
Total weight loss of sow (kg)	3.1	7.9	24.2
Creep feed intake (kg)	3.1	3.0	2.6
Litter weaning weight (kg)	63	61	52

Source: Stansbury, McGlone and Tribbie (1987).

Similar results have also been reported by Lynch (1977) in Fermoy, Ireland.

Inter-relationships between pregnancy and lactation

Feeding strategies and the consequences thereof in one part of the reproductive cycle have a large effect on the remainder of the cycle. Data in Table 5.8 illustrate one effect—the more feed that is given in pregnancy the less the sow will eat under *ad libitum* conditions in lactation. When feed intake in pregnancy decreased from 2.6 to 1.6 kg per day, feed intake in lactation was 4.7 kg per day compared with 5.9 kg per day. As a result of this, there were associated changes in P_2 measurements, which indicated substantial fat losses from the animals which were consuming less feed during lactation. In addition, the lower the feed intake during lactation, the greater the weight loss of the sows and the longer the interval between weaning and oestrus. In this study the difference was 15.3 days compared to 4.0 days for the two extremes of feed intake.

In addition to the close relationship between energy intakes in pregnancy and lactation, the protein content of the feed in pregnancy has marked effects on feed intake during lactation, as illustrated in Table 5.9. When different levels of protein were fed during pregnancy and lactation, there was a large interaction between the two in terms of feed intake. Under *ad libitum* feeding

Table 5.8 The effect of feed intake during pregnancy on feed intake and performance of the sow during lactation

Lactation	Pregnancy feeding level (kg/day)					
	2.6	2.4	2.2	2.0	1.8	1.6
Feed intake (kg/day)	4.7	4.8	5.2	5.9	6.1	5.9
Sow weight change (kg)	−15.4	−9.1	−3.1	−6.4	4.5	5.7
Change in P_2 (mm)	−2.8	−1.4	−1.5	−1.1	1.8	0.0
Interval from weaning to oestrus (days)	15.3	11.7	5.9	7.1	4.1	4.0

Source: Harker and Cole (1985).

Table 5.9 The influence of dietary crude protein level in pregnancy and lactation, on feed intake in lactation

Dietary protein concentration in pregnancy (%)	Feed intake in lactation (kg/day)	
	Dietary protein concentration in lactation (%)	
	12	18
9	4.2	6.2
13	4.8	6.5
17	5.9	6.2

Source: Mahan and Mangan (1975).

conditions, a low protein level during both pregnancy and lactation resulted in a feed intake of 4.2 kg per day, compared with 6.5 kg per day when the protein level was increased during pregnancy and lactation. However, the highest protein level during pregnancy did not increase feed intake further, and this is one of the reasons for current interest in feeding systems which provide a low protein level during pregnancy and a higher protein level during lactation.

Table 5.10 provides a summary of two experiments in which sows were given a range of feeding levels during lactation, from 33 to 67 MJ metabolisable energy per day. As a result of these large changes in feed intake there were large changes in the body weights of the sows, with considerable consequences for reproduction. Thus, only 58 percent of sows at the low level of feeding were in oestrus within 7 days compared with almost 95 percent when the sows were allowed to feed *ad libitum*. There were only small increases in these values up to 3 weeks from weaning, at which time only 70 percent of sows fed 33 MJ per day had come into oestrus compared with 100 percent at the highest level of feeding. These experiments demonstrated that feeding level during lactation not only affects the requirements of the sow but that there is also a carry-over effect onto subsequent reproductive performance.

Table 5.10 Effect of energy intake in lactation on the return to oestrus in gilts

| | ME intake (MJ/day) | | | | |
	33	42	50	59	67
Bodyweight change (kg)	−23	−19	−12	−5	−2
% of sows in oestrus by:					
7 days post weaning	58	82	88	88	95
14 days post weaning	69	90	92	94	97
21 days post weaning	69	92	92	98	99

Sources: Reese, Moser, Peo, Lewis *et al* (1982) and Nelssen (1983).

NUTRIENT REQUIREMENTS

The nutrient requirements of sows are the summation of the following individual components:

* Maintenance
* Body tissue gain (or loss)
* Conceptus growth/milk production
* Exercise
* Environment
* Wastage.

With sows kept indoors, total requirements are generally considered to be the sum of requirements for maintenance of body tissues, for body tissue gain, for growth of the conceptus during pregnancy and for milk production in lactation. In an outdoor situation, however, the animal has the opportunity to move around more, and some estimate has to be made of the energy expended due to exercise. In addition, the environment is very variable and account has to be taken of different conditions, such as hot and sunny, or wet, cold and windy weather; insulation of the ark also has an effect.

The nutrient requirements of the indoor sow are known relatively precisely, and recent estimates have been published for pregnant gilts and sows (Close, 1987). Table 5.11 illustrates how net gains in pregnancy can be targeted by adjusting daily energy intakes, taking into account the protein level in the diet and the body weight of the sow.

Table 5.11 Energy allowances for pregnant gilts and sows kept indoors

Target net gain in pregnancy (kg)	22.5		27.5		32.5	
Crude protein content of diet (%)	12	16	12	16	12	16
Bodyweight at mating (kg)	DE intake (MJ/day)					
120	24.2	21.6	25.7	22.8	27.1	24.0
160	27.3	24.2	28.7	25.6	30.1	27.2
200	30.2	28.5	31.5	30.0	32.9	31.5

Source: Close (1987).

For the outdoor sow additional considerations must be taken into account, especially exercise and environmental needs.

Exercise

Figure 5.7 illustrates that as pigs increase their walking speed, their heat production (a measure of energy expenditure) increases rapidly.

Estimates of how far a sow walks per day range from 1 to 10 kilometres. Energy expenditure has been measured with pigs on treadmills running at up to 6 kilometres per hour (Petley and Bayley, 1988). From these data it can be calculated that an additional 1 MJ of metabolisable energy per day approximately would be used for each kilometre walked (Table 5.12).

Environment

The sow's environment comprises a number of components, such as temperature, air movement, radiation, humidity, rainfall, snow, the absence or presence of bedding, shelter, shade and the opportunity to wallow. Estimates have been made of the pig's lower critical temperature, which represents the point at which the animal's metabolism begins to increase when temperatures are reduced below the zone of thermal neutrality. In groups of housed sows the lower critical temperature is 14°C, compared with 20°C for an individual animal. Below the critical temperature heat output

Figure 5.7 Heat production from pigs walking at different rates

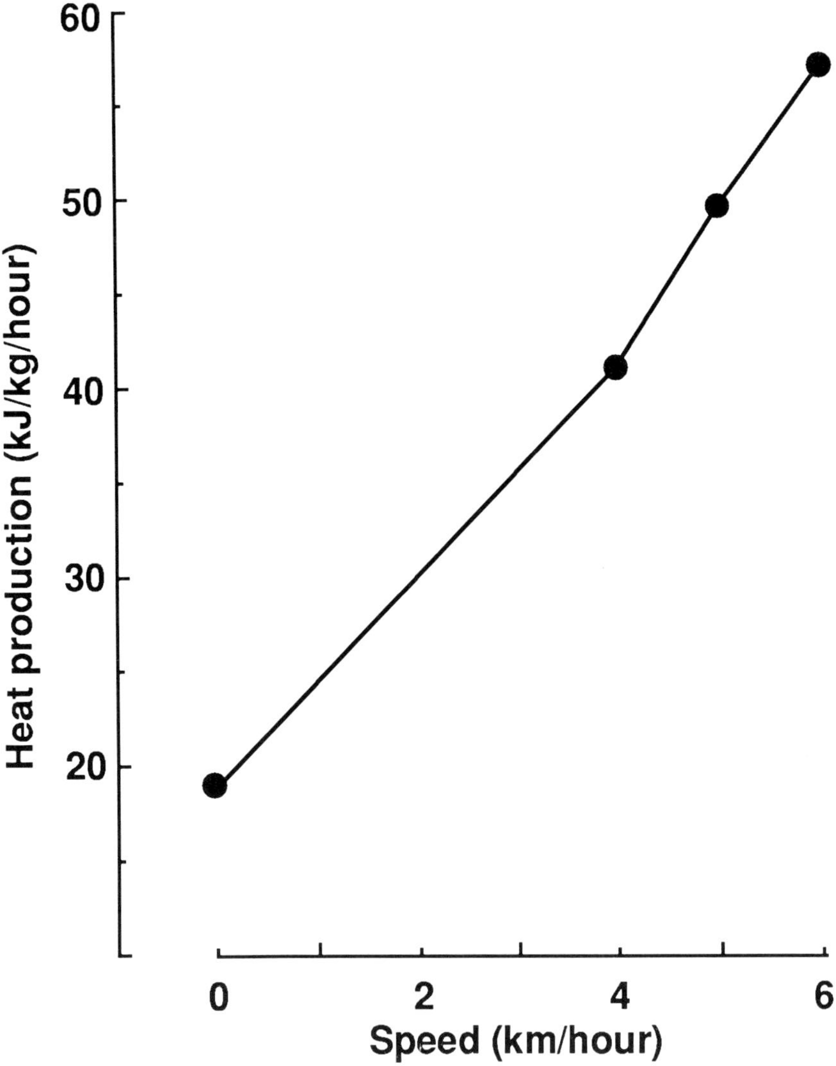

Source: Petley and Bayley (1988).

increases to a much greater extent for an individual compared with an animal in a group (see Figure 5.8), and the increase in heat output may be used to calculate the additional feed requirements of pigs in cold environments. The heat output of a 160 kg individual increases by about 0.6 MJ metabolisable

Table 5.12   Energy expenditure associated with exercise

	Additional energy required (MJ ME/day)	
	Distance walked	
Bodyweight (kg)	1 km	2 km
120	0.84	1.68
160	1.12	2.24
200	1.40	2.80

Figure 5.8   Heat production in relation to environmental temperature for pigs kept individually or in groups

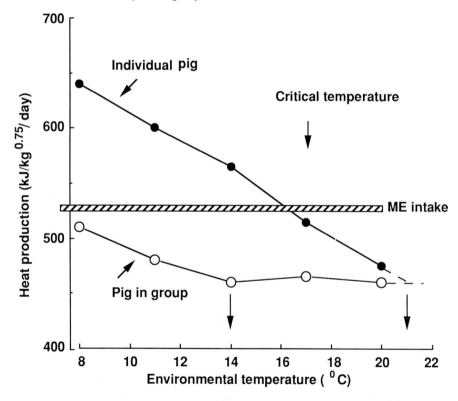

Source: Geuyen, Verhagen and Verstegen (1984).

energy per day for each 1°C below the critical temperature, and this represents the amount of extra feed energy needed to compensate for the increased heat loss. Thus, estimates of critical temperature provide a means of comparing environments within which pigs live.

Figure 5.9 models the effects of energy intake and wind speed on the lower critical temperature of sows weighing about 160 kg which live in groups, are in good body condition, and have plenty of straw bedding. As feed (energy) intake increases, the critical temperature decreases, whereas in a draughty environment the lower critical temperature increases. Critical temperature also varies with body weight, so that for pigs weighing 120 kg the critical temperature would have been higher, and for 200 kg pigs it would have been lower.

Figure 5.9 The effect of energy intake and wind speed on the lower critical temperature of 160 kg sows kept in groups, in good condition and with straw bedding

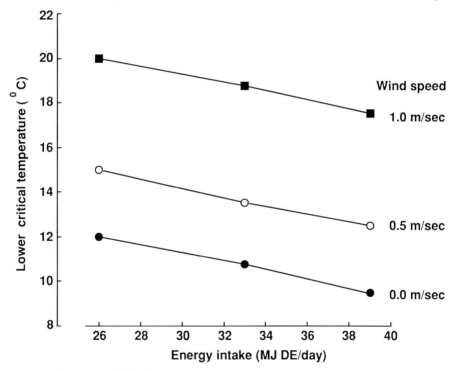

Source: Close (unpublished).

The additional energy requirements associated with the environment may be calculated. If an average critical temperature of an outdoor sow is assumed to be 15°C, Figure 5.10 indicates that there are about 1600 degree days per year when the temperature falls below 15°C; with a base temperature of 10°C, there would be only 800 degree days at the lower temperature.

Assuming pregnant animals of 160 kg, which are non-lactating for 300 days per year and hence will be fed restrictively, for each day when the temperature falls below 15°C, there is an additional requirement of 3.3 MJ metabolisable energy (Table 5.13). This represents a substantial amount of feed. For a base temperature of 10°C, using the same approach, an additional 1.6 MJ metabolisable energy per day is required (Table 5.14).

Figure 5.10   Degree days below the critical air temperature

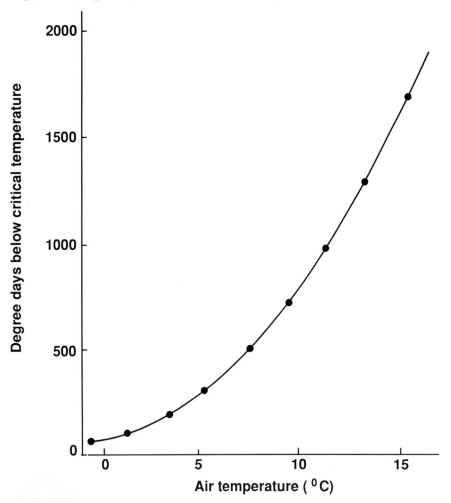

Source: Smith (1974).

Table 5.13   Additional energy requirements assuming 1600 degree days below a base
temperature of 15°C each year

Bodyweight (kg)	Additional energy requirement		
	MJ ME/day/1°C	MJ ME/year	MJ ME/day
120	0.5	784	2.6
160	0.6	976	3.3
200	0.7	1149	3.8

Table 5.14   Additional energy requirements associated with different base temperatures

Bodyweight (kg)	Additional energy requirement (MJ ME/day)	
	Base temperature (°C)	
	10	15
120	1.3	2.6
160	1.6	3.3
200	1.9	3.8

Thus outdoor sows can be fed to achieve a specific body weight gain under different environmental conditions. Table 5.15 gives the energy requirements of a gilt kept outdoors to achieve a target net body weight gain in her first parity of 35 kg, followed by gains of 27.5 kg and 20 kg in subsequent parities. Total energy requirements are the sum of the needs of animals kept indoors, plus the requirements for exercise and to compensate for the outdoor environment; the latter figure includes an allowance for feed wastage. It therefore appears that the total requirements for outdoor sows, as a mean throughout the year, are 17 percent higher than those for indoor animals, which is in agreement with the data on feed usage from the Pig Management Scheme (Chapter 1). These values are probably minimum estimates and feed requirements may be higher.

Table 5.15   Energy requirements of outdoor sows during pregnancy

Bodyweight (kg)	Net gain (kg)	Energy requirement (MJ DE/day)			
		Indoor	Exercise	Environment	Total
120	35	25.0	0.8	2.7	28.5
160	27.5	26.0	1.2	3.5	30.7
200	20	27.0	1.5	4.0	32.5

APPLICATION OF NUTRITIONAL PRINCIPLES TO FEEDING PRACTICE

In practice, diets of various energy concentrations are available and there are different types of animals in the herd. Table 5.16 gives the feeding levels

80

Table 5.16 Feed requirements of outdoor sows during pregnancy

Dietary energy concentration (MJ DE/kg)	Feed requirement (kg/day) Bodyweight (kg)		
	120	160	200
12.0	2.1 to 2.6	2.3 to 2.8	2.4 to 3.0
12.5	2.1 to 2.5	2.2 to 2.7	2.3 to 2.9
13.0	2.0 to 2.4	2.1 to 2.6	2.2 to 2.8

which would be needed for several dietary energy contents to meet the weight gain targets previously mentioned, for 3 types of sows in the herd. A range of feed intakes is given for each situation, from 10 percent lower to 10 percent higher than the mean, representing feeding levels in summer and winter respectively.

Animals kept outdoors also have the opportunity to graze, but information about the feeding value of grass for pigs is difficult to obtain. Based on data from Chambers (1987), the sow may benefit to the extent of an extra 4 kg body weight gain during pregnancy. Sows may also consume substantial quantities of bedding, especially straw, which obviously has some feeding value, but these extra bulky feeds should be treated as a bonus rather than as a part of the feeding strategy.

Figure 5.11 Feeding profile for outdoor gilts and sows

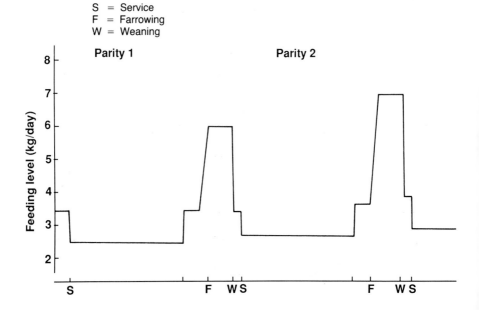

S = Service
F = Farrowing
W = Weaning

On the basis of the information provided in this paper, Figure 5.11 shows the target feeding profile which should be developed for an outdoor pig unit. The first parity gilt should be fed at a high level during oestrus to obtain as high an ovulation rate as possible, by the flushing effect. Feed intake may then be reduced, but over the last period of gestation must be increased to take account of the mobilisation of tissue in late gestation. During lactation feed intake should be as high as possible, at least 6 kg fed *ad libitum* during the first parity, with as rapid an increase as possible after farrowing. Feed intake should be decreased at weaning, but again higher levels should be fed during the weaning to mating period than in pregnancy. The feeding pattern of the first parity should then be repeated, but with feed intake at a higher level.

CONCLUSIONS

When feeding outdoor gilts, the aim is to ensure sufficient fat reserves to compensate for adverse environmental conditions. Animals should therefore be grouped at an earlier age than with indoor gilts and fed appropriately. During oestrus, nutrition should be manipulated to obtain as high a litter size as possible; husbandry and management are also important. During pregnancy, feeding strategies should minimise loss of body tissue, and during lactation feeding should be *ad libitum* to ensure minimum weight loss of the sow.

It is difficult to correct for adverse weather and other environmental conditions, but the use of bedding, insulated huts, wallows, shade, and appropriate management can minimise environmental effects. Husbandry needs should be assessed in relation to the different management and production systems practised, and stockmanship is important. As far as possible, nutritional management should aim to consider individual animals or small groups rather than the whole sow herd.

REFERENCES

Anderson, L.L. and Melampy, R.M. (1972) Factors influencing ovulation rate in the pig. In: Cole, D.J.A. (Ed) *Pig Production*. Butterworths, London, 329-366.

Chambers, J. (1987) Feeding outdoor pigs; electronic sow feeders and other methods. *The Pig Veterinary Society Proceedings*, **18**, 62-66.

Close, W.H. (1987) Some conclusions of the AFRC Working Party on the energy requirements of sows and boars. *Animal Production*, **44**, 464.

Close, W.H. and Cole, D.J.A. (1986) Some aspects of the nutritional requirements of sows: their relevance in the development of a feeding strategy. *Livestock Production Science*, **15**, 39-52.

Close, W.H., Noblet, J. and Heavens, R.P. (1985) Studies on the energy metabolism of the pregnant sow. 2. The partition and utilisation of metabolisable energy intake in pregnant and non-pregnant animals. *British Journal of Nutrition*, **53**, 267-279.

Cunningham, P.J., Naber, C.H., Zimmerman, D.R. and Peo, E.R. Jr. (1974) Influence of nutritional regime on age at puberty in gilts. *Journal of Animal Science*, **39**, 63-67.

Dutt, R.H. and Chaney, C.H. (1968) Feed intake and embryo survival in gilts. *Kentucky Agricultural Station, Progress Report No. 176*, 33-35.

Geuyen, T.P.A., Verhagen, J.M.F. and Verstegen, M.W.A. (1984) Effect of housing and temperature on metabolic rate of sows. *Animal Production*, **38**, 477-486.

Greenhalgh, J.F.D., Elsley, F.W.H., Grubb, D.A., Lightfoot, A.L., Saul, D.W., Smith, P., Walker, N., Williams, D. and Yeo, M.L. (1977) Coordinated trials on the protein requirements of sows. 1. A comparison of four levels of dietary protein in gestation and two in lactation. *Animal Production*, **24**, 307-321.

Greenhalgh, J.F.D., Baird, B., Grubb, D.A., Done, S., Lightfoot, A.L., Smith, P., Toplis, P., Walker, N., Williams, D. and Yeo, M.L. (1980) Coordinated trials on the protein requirements of sows. 2. A comparison of two levels of dietary protein in gestation and four in lactation. *Animal Production*, **30**, 395-406.

Harker, A. and Cole, D.J.A. (1985) The influence of pregnancy feeding on sow and litter performance during the first two parities. *Animal Production*, **40**, 540.

Hartog, L.A. den and van Kempen, G.J.M. (1980) Relation between nutrition and fertility in pigs. *Netherlands Journal of Agricultural Science*, **28**, 211-227.

Lynch, P.B. (1977) Effect of environmental temperature on lactating sows and their litters. *Irish Journal of Agricultural Research*, **16**, 123-130.

Mahan, D.C. and Mangan, L.T. (1975) Evaluation of the various sequences on the nutritional carry-over from gestation to lactation with first-litter sows. *Journal of Nutrition*, **105**, 1291-1298.

Meat and Livestock Commission (1983-1988) *Pig Yearbook*. Meat and Livestock Commission, Milton Keynes.

Mullan, B.P. and Close, W.H. (1989) The partition and utilisation of energy and nitrogen by sows during their first lactation. *Animal Production*, **48**, 626-627.

Nelssen, J.L. (1983) Effects of source of energy and energy intake during lactation on reproductive performance and serum hormone concentration of sows following weaning. *Ph.D. Dissertation*, University of Nebraska, Lincoln.

Petley, M.P. and Bayley, H.S. (1988) Exercise and post-exercise energy expenditure in growing pigs. *Canadian Journal of Physiology and Pharmacology*, **66**, 721-730.

Reese, D.E., Moser, B.D., Peo, E.R. Jr., Lewis, A.J., Zimmerman, D. R. Kinder, J.E. and Stroup, W.W. (1982) Influence of energy intake during lactation on the interval from weaning to first oestrus in sows. *Journal of Animal Science*, **55**, 590-598.

Smith, C.V. (1974) Farm buildings. In: Monteith, J.L. and Mount, L.E. (Eds) *Heat Loss from Animals and Man*. Butterworths, London, 345-365.

Stansbury, W.F., McGlone, J.J. and Tribble, L.F. (1987) Effects of season, floor type, air temperature and snout coolers on sow and litter performance. *Journal of Animal Science*, **65**, 1507-1513.

Toplis, P., Ginesi, M.F.J. and Wrathall, A.E. (1983) The influence of high food levels in early pregnancy on embryo survival in multiparous sows. *Animal Production*, **37**, 45-48.

Wheeler, G.E. (1987) Reproductive problems in outdoor pigs. *The Pig Veterinary Society Proceedings*, **18**, 41-61.

DISCUSSION

Nutrient requirements for pigs are calculated in terms of energy, thus different amounts of feeds of different nutrient densities are required to supply the same quantity of energy. It cannot be assumed that all sow rations contain the same level of energy, therefore feed intake and energy intake are not necessarily comparable. Some compounders formulate feeds with different energy and/or protein levels for outdoor herds compared with indoor systems, but others do not.

Many outdoor producers form groups of sows post-weaning over a period of 2 to 3 weeks, with animals mated at different times during this period. There may then be a conflict between the need to feed high levels of energy prior to ovulation and the adverse effects on embryo mortality of high feeding levels in the immediate post-conception period because sows at different stages of the oestrus/mating cycle are group-fed. The ideal would be to feed sows individually during the week following weaning, which would allow feeding levels to be applied depending on the condition of the animal, loss of weight during lactation and general productivity. However, there may also be a difference in the response of animals, which may be parity-dependent. Thus gilts and first parity sows are more sensitive to feeding than are older sows, and the results of Toplis *et al* (1983) showed that different feeding levels in the post-conception period had no effect on embryonic mortality in older sows.

There is a considerable problem in matching scientific theory and practical reality in the outdoor situation. Between the 1960s and 1980s the management of pigs became increasingly complex and intensive, but more recently there has been an increase in the number of outdoor units, which are low input systems but which necessitate good practicable management and a high level of stockmanship. Nutritional strategies must be based on simplicity.

In general, outdoor producers seem to prefer to feed larger amounts of a lower nutrient density feed because the feed has to be spread around more, for example on the ground when sows are fed in groups. This may reduce the accuracy of feeding the individual sow and may increase wastage. The important point is that if a lower density feed is used, then *larger quantities* must be fed.

The use of electronic feeders may be one way to feed individual outdoor sows more precisely, by both increasing the nutrient allowance if body condition is below average for the group and allowing animals to consume their full feed allowance in a non-competitive environment. While some producers have had problems maintaining electronic feeders in working order, outdoor production systems are now developing in which they function efficiently.

CHAPTER 6

FORMULATION, COMPOUNDING AND RAW MATERIAL USE IN FEEDS FOR OUTDOOR PIGS

P Poornan

SUMMARY

Feed formulations for pigs have changed over the last ten years, due to the increased expectation in performance of the pigs and to global changes in the cost and availability of raw materials. The widespread use of computers to formulate feeds and to control the manufacturing process means that an increasing number of raw materials may be used to provide the essential nutrients. The major advantage is that the cost of compound feeds may be kept to a minimum.

The feed formulation computer is an essential tool to enable the nutritionist to provide the nutrients required for growth and production at the least cost. However, the simple least-cost system has its pitfalls, since it is only possible to see the cheapest formula. Parametric or "what if?" formulation allows the nutritionist to see the benefits in terms of cost per unit nutrient and, using models, to estimate the cost per unit liveweight gain. This formulation process is essential in order to meet the variable nutritional requirements of the outdoor pig. A formula consists of a nutritional and a raw material specification, so that the evaluation of raw materials for use in compound feeds has to be considered from the point of view of both mill and farm. Raw material evaluation is becoming more complex and involves technical appraisal, cost, pelletability, availability, setting minimum and maximum inclusion rates, palatability, and determining storage capacities and mill requirements.

There are many practical considerations for feeding outdoor pigs, including the type of presentation, such as pellets, rolls or biscuits. Ingredients with good binding capabilities and waterproofing qualities must be used. The problems caused by competition during feeding, by sunburn, by high and low temperatures, by day length, and by soil contamination must be considered, and the feed must be formulated with these practical problems in mind.

Very little work has been published on the feed requirements of the outdoor pig. To date, it has been assumed that it is sufficient to feed more of a ration formulated for housed sows to sows kept outdoors. We believe that, due to the different feeding and behavioural patterns of outdoor pigs, they have different requirements to animals kept indoors, in terms of both nutrients and the type of raw material used. Further work needs to be done in this area to enable the outdoor producer to maximise productivity.

INTRODUCTION

In the 1920s it was common practice to feed outdoor sows on brewers wash or swill twice a day, together with green feeds such as vetches, peas, beans and clover in summer, or swedes and turnips in winter. In 1925, the Ministry of Agriculture and Fisheries published 'Pig Keeping—Bulletin No. 32', which advised pig farmers about the nutritional value of raw materials and pig feed formulations. As one would expect, the raw material list included:

> Whole milk, separated milk, whey
> Oats, barley, wheat
> Wheat middlings (wheatings), wheat bran
> Pasture grass (grazing)
> Fish meal, meat meal, blood meal
> Beans, peas, acorns
> Potatoes, sugar beet, kale
> Turnips, swedes, artichokes

Surprisingly, the raw material list also included:

> Flaked maize
> Maize gluten feed
> Palm-nut kernel cake
> Coconut cake
> Rape
> Sugar beet pulp
> Molasses
> Tapioca

Seventy years later, the outdoor pig industry is nutritionally more aware, feed formulation computers are available and animal growth models have been prepared. Nevertheless, the raw material selection to make outdoor sow feeds has remained virtually the same. Barley, wheat, wheat offals and vegetable proteins are still the main feed ingredients, although it is no longer cost-effective to feed skimmed milk to outdoor sows.

In terms of production, it was known that sows maintained in a relatively lean body condition were more prolific and more likely to breed regularly than if they were kept in a better (fat) condition. After farrowing, the diet was improved with a thick wash, comprising skimmed milk, wheat pollards and bran. Wheat offals were preferred, as barley meal was found to be more fattening and to produce less milk. Thus, even at this time a two-feed, dry sow/lactating sow system was practised.

INGREDIENTS IN COMPOUND FEEDS

Table 6.1 shows the results of a MAFF survey into the raw materials used to manufacture sow feeds. The survey is carried out on a regular basis and covers about 40 percent of the total feed production in Great Britain. It is based on typical feed formulations at a number of feed mills, so the results provide a broad, rather than a specific, indication of feed composition.

Table 6.1 Composition of compound feeds for pregnant and lactating sows, 1986 to 1988

	Composition (%)				
	1986 Jan-June	1986 July-Dec	1987 Jan-June	1987 July-Dec	1988 Jan-June
Total cereals	42	44	38	40	39
Wheat	36	31	32	34	32
Barley	5	12	5	6	6
Total cereal by-products	25	29	27	28	27
Wheat offals	23	22	24	23	23
Animal/vegetable proteins	21	17	19	18	20
Soyabean meal	12	10	11	8	9
Miscellaneous	12	11	15	13	14
Molasses	6	5	7	7	7
Manioc	—	—	1	—	—

Figure 6.1 Inclusion of cereals and protein sources in compound feeds for pregnant and lactating sows, 1981 to 1988

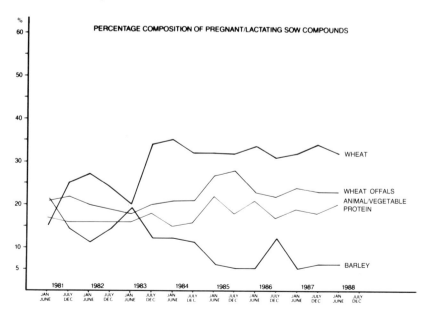

This survey shows that a sow feed typically comprises about 40 percent cereals, of which 30 to 35 percent is wheat. In the past barley was the predominant cereal, but it now makes up only 5 to 10 percent of the feed. Simple cereal by-products, such as wheat offals, account for some 25 to 30 percent of the feed. Animal and vegetable proteins such as fishmeal, meat and bone meal, soyabean meal and rapeseed meal comprise about 20 percent of the feed. Miscellaneous ingredients include fat, binders, premixes, dicalcium phosphate and salt, with molasses present at about 6 to 7 percent. Tapioca comes into formulations occasionally, when it is cost-effective.

Figure 6.1 is a graphical representation of similar data, from 1981, showing the broad groups of feed ingredients. Wheat inclusion is inversely related to that of barley; these two cereals provide the starch component of the diet, and feed formulation computer programmes generally substitute one for the other, according to cost. Wheat offal and animal protein inclusion rates are relatively constant, except when there are large changes in cereal or vegetable protein prices.

These data relate to all types of sow feeds and they are therefore representative of feeds manufactured for outdoor pigs, given that most feed compounders produce similar, if not the same, feeds for indoor and outdoor herds. It is generally assumed that outdoor sows require larger amounts of the same feed as indoor animals. By contrast, we believe that outdoor sows require a totally different feed, and for a number of years we have formulated feeds specifically for the outdoor situation.

Table 6.2 and Figure 6.2 show survey information relating to compound feeds for rearing young pigs. Total cereals comprise about 50 percent of the ingredients, with wheat 40 to 45 percent, and relatively little barley present. The inclusion of wheat offal is much lower, as diets for smaller animals tend

Table 6.2 Composition of compound feeds for rearing piglets, 1986 to 1988

	Composition (%)				
	1986 Jan-June	1986 July-Dec	1987 Jan-June	1987 July-Dec	1988 Jan-June
Total cereals	54	50	56	55	56
Wheat	44	46	42	39	44
Barley	9	3	12	15	11
Total cereal by-products	9	10	6	7	5
Wheat offals	8	7	5	5	4
Animal/vegetable proteins	32	34	32	32	31
Soyabean meal	22	25	23	22	21
Miscellaneous	6	7	6	6	7
Molasses	2	2	2	2	2
Manioc	—	—	—	—	—

Figure 6.2   Composition of compound feeds for rearing piglets, 1981 to 1988

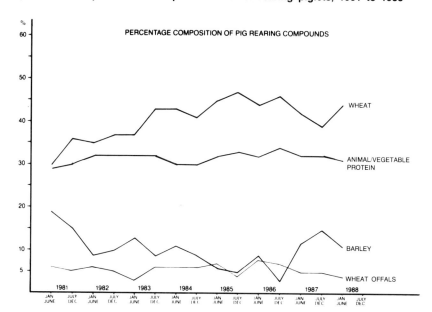

to contain less fibre and higher levels of protein and energy. The percentage inclusion of vegetable and animal proteins is therefore greater.

Figure 6.2 illustrates again that the inclusion of wheat is almost a mirror image of that of barley, while animal and vegetable protein content remains fairly constant. Over the last few years, the inclusion of wheat has tended to increase and that of barley to decrease in all pig diets.

Table 6.3 and Figure 6.3 show data for finishing compounds. In many ways these are very similar to sow diets, though certain raw materials have to be used at different inclusion rates. Total cereals form 40 to 50 percent of the ration, of which wheat is the major component. Wheat offals and vegetable proteins are fairly constant, and soyabean meal is the main component vegetable protein.

SPECIMEN FORMULATIONS

Tables 6.4 to 6.7 illustrate the use of a Linear Programming software program which is used by feed compounders to formulate diets. The nutrient values used to formulate outdoor sow rations include protein, oil, fibre, ash, lysine, methionine, threonine, digestible energy (DE), calcium, phosphorus, salt, and the ratio of energy to lysine.

Table 6.3 Composition of compound feeds for finishing pigs, 1986 to 1988

	Composition (%)				
	1986 Jan-June	1986 July-Dec	1987 Jan-June	1987 July-Dec	1988 Jan-June
Total cereals	46	46	45	44	43
Wheat	35	39	40	38	35
Barley	11	7	4	5	8
Total cereal by-products	15	18	16	15	18
Wheat offals	13	14	13	13	15
Animal/vegetable proteins	30	26	29	30	27
Soyabean meal	23	21	22	20	19
Miscellaneous	9	10	10	11	11
Molasses	4	5	5	5	5
Manioc	—	—	—	—	1

Figure 6.3 Inclusion of cereals and protein sources in pig finishing compounds, 1981 to 1988

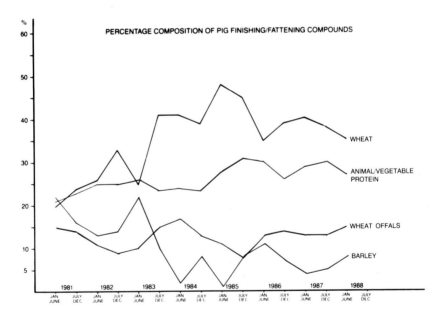

A traditional raw material specification for an outdoor sow diet is shown in Table 6.4. Wheat middlings form a maximum of about 30 percent; wheat is included at a minimum of 15 percent and a maximum of 30 percent. Barley has a minimum of 5 percent inclusion and a maximum of 25 percent. Hi-pro soyabean meal inclusion is almost without constraint, and full fat soyabeans may also be included for their high oil and energy contents. Fishmeal, molasses, blended vegetable fat, a trace mineral and vitamin premix, limestone, dicalcium phosphate, lysine and salt are also on offer for inclusion.

Table 6.4 Summary of a raw material specification for a traditional outdoor sow diet, using a linear programming package

Nutrient	Minimum	Maximum	Nutrient	Minimum	Maximum
Volume	100.00	100.00	Methionine+cystine	0.00	100.00
Dry matter	0.00	100.00	Tryptophan	0.00	100.00
Protein	16.00	16.50	Threonine	0.00	100.00
Oil	4.00	5.50	DE (pig)	13.00	100.00
Fibre	0.00	4.50	Calcium	1.00	1.10
Lysine	0.78	100.00	Phosphorus	0.60	100.00
Hardness	0.00	100.00	Salt	0.42	0.45
Methionine	0.00	100.00	Ash	0.00	6.00

Raw material	Minimum	Maximum
Wheat middlings	0.00	30.00
Wheat	15.00	30.00
Barley	5.00	25.00
Soyabean meal–dehulled	0.00	100.00
Full fat soyabeans	0.00	10.00
Chilean fish meal	2.50	10.00
Molasses and whey	5.00	5.00
High energy blended vegetable fat	0.00	1.80
Outdoor sow premix	0.25	0.25
Limestone	0.00	100.00
Di-calcium phosphate	0.00	100.00
Salt	0.00	100.00
Lysine hydrochloride	0.10	100.00

Table 6.5 illustrates the formulation of a least cost ration using the hypothetical raw material costs shown. The inclusion rate of the different ingredients is calculated to provide the required nutrients. The raw material percentages, their costs and the inclusion limits are shown. In this example, the programme would like to use more wheat middlings, more barley and less fishmeal. It suggests that there are cheaper ways of providing protein than using fishmeal, but in this instance the amino acid profile of animal protein is required. The resulting analysis and the nutrient limits of the specification (Table 6.6) show that the feed has the minimum 4 percent oil and 0.78 percent lysine required. Calcium and salt are on their maximum constraints, and phosphorus is on its minimum.

The feed compounder then has the opportunity to look at the cost sensitivity of the ration (Table 6.7), which is the effect of increasing or decreasing raw material or nutrient levels on the cost per tonne of feed. This facility is used to examine ways of reducing the cost of the feed, or to look at the cost of

Table 6.5 Summary of a least cost formulation print-out for an outdoor sow biscuit

Sensitivity		Level	Minimum	Maximum	Lower level	Unit cost	Cost difference	Upper level	Unit cost	Cost difference
Volume	(min)	100.00	100.00	100.00	99.17	1.14	−0.95	100.00	1.14	0.00
Dry matter		86.32	0.00	100.00	86.32	3.34	0.01	86.32	1.34	0.00
Protein		16.10	16.00	16.50	16.00	1.08	0.11	16.47	1.38	0.50
Oil	(min)	4.00	4.00	5.50	2.70	0.96	−1.25	4.01	0.96	0.01
Fibre		4.31	0.00	4.50	4.25	3.33	0.21	4.31	14.34	0.00
Lysine	(min)	0.78	0.78	100.00	0.77	18.16	−0.15	0.81	18.16	0.50
Hardness		5.70	0.00	100.00	5.70	4.02	0.01	5.80	2.30	0.21
Methionine		0.25	0.00	100.00	0.25	68.30	0.11	0.25	96.41	0.50
Methionine + cystine		0.49	0.00	100.00	0.49	34.10	0.11	0.50	48.19	0.50
Tryptophan		0.19	0.00	100.00	0.19	60.62	0.11	0.20	85.28	0.50
Threonine		0.53	0.00	100.00	0.53	22.58	0.11	0.55	31.78	0.50
DE (pig)		13.33	13.00	100.00	13.31	35.69	0.52	13.33	3.31	0.01
Calcium	(max)	1.10	1.00	1.10	1.00	2.09	0.21	1.12	2.09	−0.03
Phosphorus	(min)	0.60	0.60	100.00	0.57	8.68	−0.30	0.66	8.68	0.52
Salt	(max)	0.45	0.42	0.45	0.42	0.46	0.01	0.49	0.46	−0.02
Ash		5.96	0.00	6.00	5.93	0.47	0.01	6.00	12.54	0.50
Wheat	(max)	30.00	0.00	30.00	29.49	0.22	0.11	32.02	0.22	−0.44
Wheat		26.35	15.00	30.00	25.38	0.52	0.50	29.51	0.07	0.21
Barley	(max)	25.00	5.00	25.00	21.77	0.07	0.21	34.65	0.07	−0.63
Soyabean meal–dehulled		6.59	0.00	100.00	6.28	0.37	0.11	7.57	0.51	0.50
Full fat soyabeans	(rej)	—	0.00	10.00	—	—	—	8.74	0.36	3.12
Chilean fishmeal	(min)	2.50	2.50	10.00	2.35	0.89	−0.14	3.67	0.89	1.04
Molasses	(max)	5.00	5.00	5.00	5.00	0.33	0.00	6.36	0.33	−0.45
High energy blend		1.79	0.00	1.80	0.29	2.09	3.12	1.80	0.95	0.01
Outdoor sow premix	(min)	0.25	0.25	0.25	0.00	17.29	−4.32	0.25	17.28	0.00
Limestone		2.18	0.00	100.00	1.92	0.80	0.21	2.18	15.81	0.00
Di-calcium phosphate		0.19	0.00	100.00	0.10	11.22	1.04	0.52	1.56	0.52
Salt		0.06	0.00	100.00	0.03	0.46	0.01	0.06	12.09	0.00
Lysine	(min)	0.10	0.10	100.00	0.07	10.09	−0.28	0.11	10.09	0.11

improving the formulation. The nutritionist of a feed compounding company has to reach a balance between the 'ideal' formula to produce, and the realistic cost at which the feed could be sold.

It is possible to produce an excellent feed that is so expensive that it would not sell. It is also possible, in certain situations, to increase energy or protein levels at very little extra cost. Cost sensitivity allows the feed compounder to maintain a balance between the quality of the feed and its commercial aspects.

Table 6.6 Analysis of sow feed formulated according to Table 6.5

	Analysis		Limit	Programming constraints Minimum	Maximum
Volume	100.00	(min)		100.00	100.00
Dry matter	86.32			0.00	100.00
Protein	16.10			16.00	16.50
Oil	4.00	(min)		4.00	5.50
Fibre	4.31			0.00	4.50
Lysine	0.78	(min)		0.78	100.00
Hardness	5.70			0.00	100.00
Methionine	0.25			0.00	100.00
Methionine + cystine	0.49			0.00	100.00
Tryptophan	0.19			0.00	100.00
Threonine	0.53			0.00	100.00
DE (pig)	13.33			13.00	100.00
Calcium	1.10	(max)		1.00	1.10
Phosphorus	0.60	(min)		0.60	100.00
Salt	0.45	(max)		0.42	0.45
Ash	5.96			0.00	6.00

Table 6.7 Summary of cost sensitivity analysis for the feed formulated in Tables 6.5 and 6.6

Ingredients included	%	Kg	Cost	Limit	Minimum	Maximum
Wheat middlings	30.00	300.00	114.98	(max)	0.00	30.00
Wheat	26.35	263.47	123.90		15.00	30.00
Barley	25.00	250.00	119.19	(max)	5.00	25.00
Soyabean meal–dehulled	6.59	65.89	177.45		0.00	100.00
Chilean fishmeal	2.50	25.00	315.00	(min)	2.50	10.00
Molasses and whey	5.00	50.00	81.90	(max)	5.00	5.00
High energy blended vegetable fat	1.79	17.88	210.00		0.00	1.80
Outdoor sow premix	0.25	2.50	1842.75	(min)	0.25	0.25
Limestone	2.18	21.79	34.97		0.00	100.00
Di-calcium phosphate	0.19	1.91	216.30		0.00	100.00
Salt	0.06	0.56	68.25		0.00	100.00
Lysine hydrochloride	0.10	1.00	2525.25	(min)	0.10	100.00
Total	100.00	1000.00				

Table 6.8 Example of an alternative ration formulation using a 'free' raw material matrix to produce a feed with the same nutrient specification as that illustrated in Tables 6.4 to 6.7

Nutrient	Minimum	Maximum	Nutrient	Minimum	Maximum
Volume	100.00	100.00	Methionine+cystine	0.00	100.00
Dry matter	0.00	100.00	Tryptophan	0.00	100.00
Protein	16.00	16.50	Threonine	0.00	100.00
Oil	4.00	5.50	DE (pig)	13.00	100.00
Fibre	0.00	5.00	Calcium	1.00	1.10
Lysine	0.78	100.00	Phosphorus	0.60	100.00
Hardness	0.00	100.00	Salt	0.42	0.45
Methionine	0.00	100.00	Ash	0.00	6.00

Raw material	Minimum	Maximum
Wheat middlings	15.00	30.00
Crisp and potato meal	2.50	12.50
Oat feed	2.50	15.00
Wheat	0.00	60.00
Grain screenings	0.00	10.00
Maize gluten feed	0.00	5.00
Molasses and whey	5.00	5.00
Meat and bone meal	1.00	5.00
High energy blended vegetable fat	0.00	3.00
Grass meal 16%	0.00	5.00
Extracted sunflower seed meal	0.00	5.00
Rapeseed meal	0.00	17.50
Palm kernel (expeller)	0.00	5.00
Sugar beet pulp (molassed)	0.00	5.00
Cotton cake	0.00	5.00
Field beans	0.00	7.50
NIS (Nutritionally improved straw)	1.00	5.00
Outdoor sow premix	0.25	0.25
Limestone	0.00	100.00
Di-calcium phosphate	0.00	100.00
Salt	0.00	100.00
Lysine hydrochloride	0.00	100.00

Ingredients included	%	Kg	Cost	Limit	Minimum	Maximum
Wheat middlings	15.00	150.00	114.98	(min)	15.00	30.00
Crisp and potato meal	7.11	71.14	119.00		2.50	12.50
Oat feed	2.50	25.00	67.00	(min)	2.50	15.00
Wheat	42.31	423.11	123.90		0.00	60.00
Molasses and whey	5.00	50.00	81.90	(max)	5.00	5.00
Meat and bone meal	1.48	14.77	185.00		1.00	5.00
High energy blended vegetable fat	2.49	24.93	210.00		0.00	3.00
Rapeseed meal	17.50	175.00	105.00	(max)	0.00	17.50
Palm kernel (expeller)	0.59	5.88	116.00		0.00	5.00
Field beans	3.06	30.63	120.09		0.00	7.50
NIS	1.00	10.00	71.40	(min)	1.00	5.00
Outdoor sow premix	0.25	2.50	1842.75	(min)	0.25	0.25
Limestone	1.63	16.25	34.97		0.00	100.00
Salt	0.08	0.79	68.25		0.00	100.00
Total	100.00	1000.00				

Table 6.7 shows the cost sensitivity of the nutrients, listing the lower and upper levels of linear cost change, and the effect on cost per tonne in moving from the current level to the new level. For example, oil is currently set to a minimum of 4 percent, but if this is decreased to 2.6994 percent, than the saving would be £1.246. However, this potential saving has to be weighed against the cost of producing new labels, informing the sales staff, and re-evaluating the legal declarations, the changes in formulation, raw material usage, pellet quality, feed performance and the farmers viewpoint. Alternatively, if the oil was increased to 4.0017 percent, then the increased cost would be £0.011.

As already stated, the formulation shown is a very traditional diet—one of Lys Mill's better-selling sow biscuits. But rapeseed meal, meat and bone meal, and a number of other raw materials have not been offered.

Table 6.8 illustrates another formulation for exactly the same nutrient specification, using a 'free' raw material matrix. The two diets would have exactly the same declaration on the label, and the same statement of energy content, but the ingredients would be totally different. This is purely an example to show that current label declarations give no indication of the quality of the ingredients used to make the feed. The second specimen feed contains 17.5 percent rapeseed meal, which is more than most (but not all) compounders would include; 5 or 7.5 percent would be the normal maximum. A large amount of wheat is included, as well as oatfeed, crisp and potato meal and 'nutritionally-improved' straw.

These two formulations demonstrate that a farmer needs to know the actual raw materials used in order to assess the potential performance of a feed. Values for the chemical analysis of oil, protein, fibre, ash and vitamins mean little; they give no indication of the origin of the 13 megajoules (MJ) of energy per kg dry matter (DM), which may be derived from, for example, starch, sugar or fat, and no indication of the digestibility and bioavailability of the nutrients in the feed.

RAW MATERIALS

There are numerous raw materials that a feed compounder can use and the diets shown are only examples. There are advantages and disadvantages to all raw materials, and the feed compounder and farmer must judge whether it is justifiable and worthwhile to use particular raw materials.

Wheat middlings are generally considered by compounders as a very cost-effective raw material and they are normally used in outdoor sow diets at some 20 to 30 percent. The protein content, which is about 15 to 16 percent, is of good quality and contains a relatively high level of lysine. Middlings are a good source of fibre and have the benefit of reducing problems of constipation.

Oat feed is a low protein, high fibre raw material which can be difficult to handle and pellet. It can be used to increase the fibre level and to reduce the energy level in the feed, as its digestibility is relatively low. It may be useful at low inclusion rates when formulating dry sow feeds.

Crisp and potato meal is a waste product from the crisp industry, and is a mixture of crisps, potato rind, shredded wheat and biscuit meal. It has high starch and oil contents and can be used to replace some of the cereals in the feed. This raw material is highly digestible but the salt level is variable, and it must therefore be used with care.

Tapioca (also known as mannioc or cassava) is a cereal replacer sometimes used, when it is sufficiently cheap, to replace wheat and barley. It is low in protein and oil, but high in starch. Tapioca contains a number of anti-nutritive factors, such as thiocynates and oxalates, the levels of which are related to production and processing methods such as sun-drying. There are problems in using this raw material and it has been shown to cause ulcers in small piglets. Tapioca has been included at 30 to 35 percent in sow diets, but it must be used with caution.

Barley and wheat are well-established cereal grains, and their quality determines, to a large extent, the nutritive and pellet quality of the feed. If these cereals are not finely ground when making outdoor sow biscuits or rolls, the grain can swell and break open the pellet. Barley may be used freely in outdoor sow diets, but wheat should be restricted to about 50 percent of the feed.

Maize is low in protein and deficient in certain amino acids, but it is a very good source of energy and unsaturated fats. It is, however, now too expensive to be included in sow feeds.

Maize gluten feed, a by-product from the starch industry, is a useful raw material for outdoor sows. There are a number of grades of maize gluten feed and the protein content may vary from around 20 percent to 30 percent of the DM; a 70 percent protein maize gluten meal is also available. Maize gluten can be unpalatable, and changes in its inclusion rate must be kept to a minimum. Maize gluten feed must not be confused with maize gluten replacer, which is a mixture of raw materials mainly used in ruminant feeds.

Molasses, as a term, now includes over a dozen different types of products. Cane molasses and beet molasses are the primary ingredients, which can be mixed with a number of other raw materials such as whey and fermentation solubles. Molasses mixed with whey is a useful feed ingredient for pigs. It provides additional nutrients, in addition to acting as a binder to help form better quality pellets; the whey also makes the molasses slightly less viscous than cane molasses.

Fishmeal is a traditional raw material with a high protein content and, being an animal protein source, it provides an amino acid profile similar to that required by the pig. There are many sources of this material, and fishmeals broadly fall into the categories of *whitefish, herring* and *anchovy* fishmeals. The digestible energy content for pigs depends on the type of fishmeal, the oil and the protein content. Fishmeal is also a good supply of bioavailable minerals and provides salt, calcium and phosphorus.

Blended vegetable fats are formulated in much the same way as compound feeds. The feed compounder can therefore specify the ingredients, nutritional quality (fatty acid profile, linoleic acid content etc) and purity of the fat. Tallow may be considered as a constituent, but since most compounders have only one source of fat for all the feeds they produce, they are now wary of using tallow due to potential problems with recycling animal by-products.

Grass meal is similar in some ways to wheat middlings. It may be used to dilute the energy content of the diet, and the high fibre level can help to produce a low density, low energy, bulky-type of feed. It is generally of relatively low digestibility, and tends to be digested in the hind-gut. Grass meal can be used at up to 7.5 percent of the feed.

Rice bran is generally of two types, namely high oil rice bran and oil-extracted rice bran. The fat in high oil bran turns rancid very quickly and, if fed to animals, can have several detrimental effects. Rancidity adversely affects palatability, the quality of the oil and energy levels; it can also lead to the production of indigestible and potentially toxic components. Rice bran which is going rancid can also destroy the fat-soluble vitamins in the feed, and can trigger the oxidation of other fat components in the feed.

Extracted rice bran obviously does not have the problem of oil rancidity, but it is commonly adulterated with limestone or other materials at the country of origin. It is also commonly infested with insects or their larvae, which can cause a number of problems in the performance of the feed and at the feedmill. Extracted rice bran has a very shiny surface and can be difficult to pellet. In view of all the problems associated with this raw material, many compounders refuse to use it.

Rapeseed meal, like other oilseed meals, contains a number of antinutritive factors: erucic acid, glucosinolates, sinapine and tannins. Compounders must ensure that these factors are kept at low levels in the feed by ensuring that the meal has been properly cooked or processed, and by limiting its inclusion in feeds. The double zero or double low rapeseed meals that are now available may be used at slightly higher inclusion rates. The fact that outdoor sows are generally under more environmental stress suggests that adverse factors in the feed, such as the glucosinolates which interfere with the control of metabolic rate, may have more effect than in feeds for sows kept indoors.

PALATABILITY

An important factor about a feed is whether the sow will eat it. The common, relatively unpalatable raw materials are maize gluten feed, rapeseed meal, sunflower seed meal and palm kernal meal, whereas molasses, barley, full-fat soyabeans and fishmeal are generally eaten readily. Nutritionists have to balance one ingredient against another to decide if the feed will be palatable.

PHYSICAL PROPERTIES OF FEEDS

Form of feed

Outdoor sows may be fed *pencils*, *rolls* or *biscuits*, and the latter may be referred to as *chunks*, *jumbos*, *cobs* or *chilvers* (Plates 6.1, 6.2 and 6.3). Pencils are not commonly used in the outdoor situation because they are small, they can easily be trampled into the mud, and they can be eaten by birds. Rolls or biscuits are more generally fed, with rolls being used when the feed is blown into a bin. Biscuits are probably the most common form of feed because they reduce wastage and trampling losses.

Pelletability

Feed compounders have attempted to put a hardness factor into the feed formulation matrix because certain raw materials make a better pellet than

Plate 6.1 Sow pencils (8 mm diameter)

SOW PENCILS 8mm

Plate 6.2   Sow rolls (16 mm diameter)

Plate 6.3   Sow biscuits

others. Very few of these systems actually work in practice, and they only increase costs. The major problem is that combinations of certain raw materials make a better or poorer pellet than the individual ingredient characteristics would indicate.

The exact formulation can have dramatic effects on pelletability; high starch levels tend to make a better pellet, while added fat reduces pellet quality. Long fibrous materials, such as grass meal, can help to 'reinforce' a pellet at low inclusion rates, but can cause pellets to break apart at high inclusion rates. The largest influence on pelletability, however, is the production and conditioning process itself.

Water resistance

The water-proofing of feeds for outdoor sows is important, as feed is often spread in the mud and the pigs have to find it. Feed may also be stored under unfavourable environmental conditions in bins and in barns. Some raw materials contain a certain amount of inherent oil which aids the water-proofing of the final product. The inclusion of additional fat can help to waterproof the feed but it adversely affects pelletability, so a balance between these two properties is needed. Fat-coating the pellet or biscuit is a possibility but, if not done correctly, the fat sticks on the outside and soil ingestion by the sow increases markedly. The extra soil ingestion decreases appetite and adversely affects the performance of the outdoor sow. Certain raw materials absorb water and swell faster than others, and if used inappropriately would break up the biscuit or roll, thus the choice of raw materials affects pellet quality.

CONCLUSIONS

As already noted, very little work has been published on feeding the outdoor sow. Studies are currently being carried out relating to the chemical composition and nutritive value of feeds for the outdoor sow. We at Lys Mill believe that, compared to animals kept indoors, outdoor pigs have different nutritional requirements, bearing in mind factors such as the environmental situation and level of soil ingestion; that they have different appetites and feeding habits; and that they can probably make better use of hind gut fermentation to enable certain raw materials to be fed that would be inappropriate for the indoor pig.

DISCUSSION

Lys Mill formulations for feeds for outdoor sows have a maximum but not a minimum constraint for fibre because animals kept outdoors generally have access to relatively large amounts of fibre from their environment, such as from straw bedding or from grass. Thus there appears to be little need for fibre in sow diets during lactation. However, there may be a niche for a high fibre, relatively low nutrient density feed post-weaning and during the gestation period, so that the dry matter intake of sows is not markedly reduced at weaning, to under 50 percent of that during lactation.

Sugar beet pulp, a useful source of digestible fibre, seems to be currently under-utilised in pig diets, being rarely included in formulations. It tends to be regarded more as a feed for ruminants, but in the future it may assume more importance in pig diets, especially for outdoor sows. Results from recent research trials should help to establish its role as an ingredient in pig feeds.

Nutrient supply, including fibre, from non-feed sources is highly variable. In some instances the environment provides little more than a limited amount of fibre from straw bedding, but in other situations most or all of the requirements of dry sows can be met by good grassland.

Feed compounders are increasingly providing more information on the ingredient and nutrient composition of pig feeds, and this is likely to continue in the future. This is the only means by which the value of different feeds can be compared. Some compounders are likely to adopt a totally open policy on feed declaration, whereas others will probably indicate groups of ingredients which may be included or which are excluded. Most compounders formulate feeds using retrospective raw material analyses, taking into account average analyses from the previous month. Most cereals which Lys Mill use are purchased on contract from farmers and are of a specified quality.

CHAPTER 7

ALTERNATIVE FEEDS FOR OUTDOOR PIGS

D H Machin

SUMMARY

There is considerable interest in reducing feed costs by using cheaper feeds and by improving the efficiency of feed use. Within the indoor pig industry major steps have been taken in both areas. By contrast, few changes have taken place with outdoor pig herds and, although alternative feeds have been used for centuries, little published information is available on feeding them to pigs kept outdoors. Mature pigs have the potential to digest cellulose virtually completely, but factors which influence the ability of animals to utilise fibrous feeds in practice include: age of the animal, previous experience of fibrous feeds, particle size of feeds, presence of anti-nutritive factors, the balance and concentration of nutrients, the degree of lignification of the fibre, the presence of antibiotics in the diet and, probably, genetic characteristics of the animal.

Several factors must be taken into account when considering the use of alternative feeds, the most important generally being their nutrient content and the nutrient requirements of the animal, which must be met within a finite appetite capacity. Feeds must, however, also be cost-effective sources of nutrients. Diets formulated for dry and lactating sows can be very different; for example there is considerable scope for including low nutrient density, fibrous feeds in dry sow rations. A number of such feeds may be suitable, including maize and grass silages, grazed grass, cabbages, swedes, turnips, fodder beet and potatoes. Economic assessments of production costs and yields per hectare suggest that these are cost-effective feeds per unit of dry matter and energy. The few trials which have been carried out with outdoor pigs have involved small numbers of animals but results were promising for systems based on rotational grazing, fodder beet and maize silage. Means of overcoming some of the remaining problems associated with feeding alternative feeds should be considered further—a possibility may be a complete diet system. In addition to economic advantages, alternative feeds for dry sows may have other benefits in terms of behaviour and welfare, increased milk fat production, increased litter size and piglet weight, a decreased lower critical temperature, improved health, reduced water consumption and an increased appetite in lactation.

INTRODUCTION

Feed accounts for up to 75 percent of the cost of pig production, and there has been considerable interest in reducing feed costs both by using cheaper feeds and by improving the efficiency of feed use. Nutrition and environment are closely related and so must be considered together in this context.

Major steps have been taken within the indoor pig industry to utilise by-product feeds, particularly liquid feeds, and to improve environmental conditions through housing design to achieve a 'nutrient sparing' effect. By contrast, there have been few widespread changes during recent years in either approaches to feeding outdoor pigs or the modification of environmental conditions.

As with other aspects of outdoor pig production, very few data have been published which relate specifically to feeding animals kept outdoors. Although considerable work on feeding alternative feeds to pigs has been carried out for centuries, and the alternative feeds discussed in this paper have been fed on outdoor pig units, feed consumption and animal performance are not well-documented. Thus, the role of alternative feeds in outdoor pig production must be assessed from first principles.

Outdoor pig production has particular advantages compared to the indoor situation when considering the development of lower cost feeding systems, in particular relating to the cost of establishment. Most interest lies in gestating and lactating sows, as piglets are generally reared indoors after weaning.

METHODS OF REDUCING FEEDING COSTS

With any production system the four main means of reducing the cost of feeding pigs are to:

1. *Increase the efficiency of nutrient usage* by careful balancing of the diet so that nutrients complement each other.

2. *Maximise the reproductive performance of sows*, to minimise maintenance requirements.

3. *Maintain the environment within the thermoneutral zone of the animal*, if possible. This is rarely achieved within arks in outdoor production systems and research is required to improve the housing of outdoor pigs.

4. *Use lower-cost feed ingredients*, which are often of a lower nutrient density than conventional feed ingredients for pigs.

FACTORS INFLUENCING THE USE OF LOW NUTRIENT DENSITY FEED INGREDIENTS

Several factors must be taken into account - when considering whether alternative feeds may be used to replace conventional feed ingredients in rations for pigs. The most important of these are:

* Voluntary feed intake;
* Nutrient requirements of the pig;
* Concentrations of nutrients in the low energy density feed(s) under consideration;
* Cost per unit of nutrient;
* Nutrient spectrum of other feeds available to balance the main feed(s).

The use of feed ingredients is generally governed primarily by their nutrient content. The appetite of a pig has a physical limit and it is therefore vital, to achieve a given level of performance, that nutrient requirements can be met within the voluntary dry matter (DM) intake capacity of the animal. The choice of alternative feed(s) must allow the minimum required nutrient density of the overall ration to be achieved. In addition, other feeds must be available to provide nutrients which are relatively deficient in the low density feed(s).

The cost per unit of nutrient in available feeds is the main factor influencing the economics of using alternative feeds. While the initial criterion when choosing feeds is that nutrient requirements must be met within appetite limits, for profitable pig production it is equally important that cost-effective feeds are used.

VOLUNTARY FEED INTAKE AND NUTRIENT REQUIREMENTS OF DRY AND LACTATING SOWS

Voluntary feed intake and nutrient requirements are given in Table 7.1. Although a typical appetite capacity of 6 kg dry matter per day is indicated, estimates of the feed intake of lactating sows vary widely, from 4.5 to 10 kg dry matter per day.

Table 7.1 Voluntary feed intake and nutrient requirements of dry and lactating sows

Requirements[1]	Dry sow	Lactating sow[2]
Digestible energy (MJ/day)	30	70
Crude protein (g/day)	140	850
Lysine (available) (g/day)	8.5	28
Voluntary feed intake (kg DM/day)	4	6

[1]Values are approximate and will vary with individual animals.

[2]Sow with 10 piglets and assuming 10 kg weight loss over lactation.

It can clearly be seen that the energy, protein and lysine requirements of lactating sows are much greater than those of dry sows. By contrast, the relative increase in appetite during lactation is considerably smaller.

Minimum nutrient concentrations in the rations of dry and lactating sows are given in Table 7.2. The required concentrations of digestible energy (DE) in the diets of dry and lactating sows, based on the data in Table 7.1, are 7.5 and 11.7 Megajoules (MJ) digestible energy per kg dry matter respectively. Protein levels should be at least 35 and 142 g crude protein per kg dry matter, respectively, for dry and lactating animals, and lysine levels 2.1 and 4.7 g per kg dry matter.

Table 7.2 Minimum available nutrient concentrations to meet dry and lactating sow requirements within voluntary feed intake constraints

	Dry sow	Lactating sow
Digestible energy (MJ/kg)	7.5	11.7
Crude protein (g/kg)	35	140
Available lysine (g/kg)	2.1	4.7

Very different diets could be fed to dry and lactating animals, which indicates that in particular there is considerable scope for including low nutrient density feeds in rations for dry sows. However, small quantities of high energy feeds are usually given to dry sows, and this method of feeding generally results in discontented animals. Lower density diets for dry sows thus have potential benefits both in economic terms and from welfare considerations.

POTENTIAL FEEDS

Feeds which may particularly suit the outdoor sow include potatoes, fodder beet, swede, turnip, cabbage, grass silage and maize silage, which all have an acceptable nutritional value. The outdoor sow could be fed whichever is available, if the nutrient costs are right and if there are no problems of toxicity or palatability. Table 7.3 shows a range of examples of feeds, together with their composition and nutritional values. Information on potatoes, fodder beet, swede and cabbage is from work by Livingstone (1988), who corrected the digestible energy value for methane production. The energy value of maize silage was determined by ADAS and may be an under-estimate (Carlisle and Mitchell, 1984).

It is thus possible to supply most of the nutrient requirements to the dry sow with bulky feeds of these particular types and, if suitably supplemented, they could also provide a large part of the nutrient requirements of the lactating sow. Work on this topic was carried out principally at two locations, the Rowett Research Institute in Aberdeen and ADAS, in association with the

Table 7.3   Dry matter, digestible energy, crude protein and lysine contents of a range of alternative feed ingredients

Raw material	Dry matter (g/kg)	Digestible energy (MJ/kg DM)	Crude protein (g/kg DM)	Available lysine (g/kg DM)
Potato (raw)	210	10.8	109	3.8
Fodder beet	151	11.4	67	1.4
Swede/turnip	100	11.8	90	3.0
Cabbage	85	10.1	230	5.4
Grass silage	380	10.9	174	6.7
Maize silage	230	8.7	83	2.8

Source: Compiled from Carlisle and Mitchell (1984), and using corrected digestible energy values of Livingstone (1988).

Dorset College of Agriculture and Dalgety Agriculture Ltd, in the south-west of England.

UTILISATION OF ALTERNATIVE FEEDS

It has been shown fairly recently that the mature pig is able to digest fibre, particularly cellulose, in a similar way to ruminants. Although farmers world-wide have been feeding fibrous products for many years, it was felt in the UK that research on this topic was not required. Only comparatively recently has there been felt to be a need to establish how the sow is actually able to use fibrous feeds, and how efficient she is in carrying this out.

Digestion and absorption of enzymatically digestible components takes place in the upper digestive tract, while bacterial fermentation occurs in the large intestine. Here, only volatile fatty acids and water can be readily absorbed.

The factors that affect the ability of pigs to utilise fibrous feeds are:

* **Age.** As a sow or a pig gets older, its ability to digest fibre and fibrous feeds increases. From about 50 to 60 kg liveweight onwards feeding fibrous feeds becomes cost-effective. Low nutrient density feeds such as fodder beet may often be fed at a lower liveweight, although most nutrients would be obtained from the non-fibrous components in the diet.

* **Previous experience of fibrous feeds** helps feed utilisation, as the digestive tract has to adapt to fibre digestion. There is some indication that the earlier in life that animals are given fibrous feeds the better able they are to utilize them later in life.

* **Particle size of feeds.** Feeds with a small particle size usually have a fast rate of passage through the digestive tract, which offers less

opportunity for fibre digestion. Therefore it is important that fibrous feeds have a relatively large particle size.

* **The presence of anti-nutritive factors** can impair the utilisation of any type of feed.

* **The balance of nutrients** in the diet can similarly influence the utilisation of fibrous and non-fibrous materials.

* **The concentration of nutrients in the feed** is important, since they must be capable of meeting the pig's requirements within its voluntary feed intake.

* **The degree of lignification of fibre and the presence of other non-cellulose and fibre components in the diet**, such as waxes. Where there is very low lignification of the crude fibre components in the diet, cellulose digestibility can be very high, up to 100 percent. With high levels of lignification in, for example, mature grasses, digestion of cellulose is quite low.

* **The presence of antibiotics** in the diet reduces the ability of the sow to ferment fibre. Further, comparatively recent work (Varel, 1987) has shown that the microorganisms in the large intestine of the sow are largely identical to those in the ruminant or in the horse.

* **Genetic characteristics of the animal** appear to be an important factor. Work with lean and obese pigs has shown that lean pigs seem to be able to use volatile fatty acids, which are the products of fermentation of cellulose, better than obese pigs (Pond, 1987).

There are thus many factors which influence the use of fibrous feeds in the diet. It is, however, clear that the sow has the potential to use materials of this type. The digestibility of cellulose in unlignified feeds may approach 100 percent, and up to 30 percent of energy intake may be derived from volatile fatty acids.

RESEARCH INTO THE PRACTICAL USE OF ALTERNATIVE FEEDS

Limited data exist on the practical use of many alternative feeds; much of the information that is available was provided from small-scale trials carried out by ADAS in the South West Region. The work was of a pragmatic type, with a small number of animals, but it provides the only information which is available on the use of these feeds. The fact that animal performance was quite acceptable, although the number of pigs was small, indicates that there is considerable potential to use alternative feeds.

Rotational grazing

In one trial, at the Dorset College of Agriculture, animals were rotated around a grazing system, using an electronic sow feeder with balancer feed designed by Dalgety Agriculture Ltd (Chambers, 1987). Although it appears that sows can obtain a considerable amount of nutrients from grassland, this rarely happens; in the trial the grazing system was designed to maintain grass on the field by moving the animals before they started to poach or to overgraze the pasture.

In this trial 2 groups of pigs were fed 2 levels of compound feed, and their performance is shown in Table 7.4. The aim with the first group, fed 1 kg compound per day, was to obtain a weight gain over a breeding cycle of 10 to 15 kg, which was achieved. Slightly lighter animals in Group 2 were fed 2 kg feed per day and their higher weight gain resulted in them being slightly overweight. In this particular trial, despite an apparent scarcity of grass, levels of performance were quite acceptable.

It must be emphasised that this work was not replicated. It was a 'look-see' study, which indicated the *potential* of grass as a feed.

Table 7.4 Performance of sows on a rotational grazing system, with a balancer feed provided from an electronic sow feeder

	Group 1	Group 2
Sow weight initially	>200 kg	<200 kg
Intake of balancer (kg/day)	1	2
Mean weight gain of sow (kg)	14.5	24.7
Mean number born alive	10.5	9.9
Mean number born dead	0.7	0.9
Post-weaning days to service	13.6	12.7

Source: Chambers (1987).

Fodder beet

Work by ADAS in conjunction with the Dorset College of Agriculture (Chambers, 1987) has also looked at the strip feeding of fodder beet, again using electronic sow feeders to supply 3 different levels of balancer compound (1.0, 1.5 and 2.0 kg per day). The strip grazing was adjusted so that fodder beet intakes were approximately 17 kg per head per day.

Sow numbers were small, from a statistical point of view, but the results given in Table 7.5 show that the level of gain required over a reproductive cycle was actually exceeded in all groups. However, the numbers of piglets born alive appeared slightly low in the group receiving only 1 kg balancer (Group 3), although statistical significance is difficult to determine with the small numbers of animals. The general impression is that this could be a potentially effective way of feeding outdoor pigs.

Table 7.5 Performance of sows strip grazing fodder beet*, with three levels of balancer provided from an electronic sow feeder

	Group 1	Group 2	Group 3
Compound balancer (kg/day)	2	1.5	1.0
Sow weight at start (kg)	187	221	241
Number of sows per group	46	22	21
Mean weight gain of sow (kg)	31	20	16
Mean number born alive	11.0	11.8	9.4
Mean number born dead	0.8	1.2	1.6
Percentage failed to farrow	0.5	0.18	Nil

* Grazing controlled—approximately 17 kg fodder beet eaten per day.

Source: Chambers (1987).

Maize silage

A further ADAS trial involved feeding maize silage (Carlisle and Mitchell, 1984). Again, small numbers of sows were involved and the approach was a 'look-see' study. Sows were fed 1 kg of balancer with approximately 11.8 kg maize silage, which was fed once a day in troughs made from tractor tyres, until 3 to 4 weeks pre-farrowing. The performance of sows fed in this way was compared with others fed compound feeds.

Results are given in Table 7.6. The estimated energy content of the maize silage appeared to be low compared to the actual feeding value, since the pigs became overweight. Reproductive performance was good.

Table 7.6 Performance of dry sows fed on maize silage, with a balancer provided from an electronic sow feeder

	Group 1	Group 2
Sow weight at start (kg)	171	184
Number of sows in group	9	16
Balancer meal (kg/day)	1	—
Sow cobs (kg/day)	—	2.7
Maize silage (kg DM/day)	11.8	—
Mean sow weight gain (kg)	32	34
Mean number born alive	12.0	11.4
Mean number born dead	nil	0.15

Electronic sow feeders

On occasions during these trials there were problems with the electronic sow feeders and, in some cases, the sows did not enter the feeder to get the balancer. In particular, in the fodder beet trial the sows often preferred to eat fodder beet alone. There were also problems in maintaining the sow feeders outside, but these have now been largely overcome. Thus, it appears that the use of electronic feeders together with alternative feeds may have a future in outdoor pig production.

YIELDS OF DRY MATTER AND ENERGY FROM ALTERNATIVE FEEDS

Table 7.7 lists the yields per hectare and costs of production of a range of alternative feeds, using values from Nix (1989) and ADAS (1984, 1986), together with digestibility values from the work of Livingstone and Fowler (1984). Yields of fresh material, dry matter and digestible energy are given, together with variable costs per hectare and per unit of digestible energy. Costs do not include harvesting or storage.

Table 7.7   Typical yields of alternative feeds

	Yield (tonnes/ha)	DM content (g/kg)	DM yield (tonnes/ha)	Yield of DE* ('000MJ/ha)	Variable costs** (£/ha)	Variable costs (pence/ MJ DE)
Potatoes[1]	37.5	210	7.9	85	1385	1.63
Fodder beet[2]	75	180	13.5	153	285	0.18
Swede turnip[3]	69	96	6.6	78	140	0.18
Grass silage[3]	55	200	11.1	121	120	0.10
Cabbage[1]	90	85	7.7	78	270	0.35
Maize silage[3]	50	250	12.6	110	180	0.16
Winter barley[3]	5.65	860	4.9	68	195	0.29
Winter wheat[3]	6.75	860	5.8	91	230	0.25

* Corrected for fermentation in the large intestine using data of Livingstone and Fowler (1984).
** These values do not include harvesting or storage costs.

Sources: [1]Nix (1989); [2]ADAS (1986); [3]ADAS (1984).

Yields of digestible energy per hectare have been corrected to allow for the losses due to fermentation that take place in the large intestine of a sow. Calculated values per hectare are 153000 MJ for fodder beet, 85000 MJ for potatoes, 78000 MJ for swedes or turnips, 121000 MJ for grass silage, 78000 MJ for cabbage, 110000 MJ for maize silage, 68000 MJ for winter barley and 91000 MJ for winter wheat. Clearly, per unit of area, very high levels of energy can be produced from each crop.

When making silage, some of the energy component of the original material is lost together with some of the available protein during fermentation, which must be taken into account if comparing silage with the original fresh material. At certain times of the year, chopped fresh maize or chopped fresh grass can be fed to sows (chopped maize from the field is a common feed overseas) and these will have higher nutritional values than silages.

The cost of producing these crops per MJ of digestible energy is also shown in Table 7.7. This is a useful basis to compare different raw materials.

FEEDING SYSTEMS FOR ALTERNATIVE FEEDS

From an economic point of view, in terms of yields per unit of land there appears to be considerable potential for using forage crops for outdoor pigs. While a number of pig producers are practising the types of system outlined, many consider that although the principle is sound making these systems work in practice is much more difficult. There have been problems using electronic sow feeders, in particular getting animals to eat the balancer, whilst pigs folded on fodder beet have become overweight.

One way of solving some of the problems would be to use a complete feeding system similar to that used for dairy cows. A complete feed, including balancer, could be prepared from a range of materials and fed *ad libitum* to animals in any location—indoors or outdoors. However, if this approach is to be used economic aspects in each situation must be carefully assessed.

Systems involving grazing or folding could also be considered, or even lifting root crops and dropping them across the paddocks, using the electronic sow feeder to provide a supplement. Further work is needed to define potential systems more closely.

OTHER BENEFITS OF ALTERNATIVE FEEDS

Other benefits of feeding low nutrient dense, fibrous feeds noted in the literature (Lee and Close, 1987) include:

*   **Behaviour and welfare**. Stereotyped behaviour has been reduced, lying time increased and the sows appeared slightly more content when bulky feeds have been fed.

*   **Increased fat content in milk**. Interestingly, the response of the sow is similar to that of a ruminant, as diets with a fibre source increase acetate production. This appears to increase the fat content in milk, which is beneficial to the piglets.

*   **Increased litter size and weight**. There have been suggestions that litter size and birth weight of piglets may be higher with a bulky type of diet.

* **Decreased lower critical temperature**. Because of the heat of fermentation, the sow appears to have a lower critical temperature when consuming fibrous feeds, possibly due to the heat that the sow herself generates from fermentation.

* **Improved health**. There have been suggestions of improved health, particularly in the digestive tract, through using diets containing bulky feed materials. For example, ulceration of the stomach may be reduced due to the feed having larger particles.

* **Reduced water consumption**. Many alternative feeds have relatively low dry matter contents, thus there is often less need for the animal to consume water from drinkers.

* **Increased appetite in lactation**. One of the most important benefits of bulky diets is that, by having the stomach stretched during gestation, sows appear to have an increased appetite during lactation, thus reducing weight loss. Conversely, high nutrient density feeds would reduce appetite during lactation.

CONCLUSIONS

* Low nutrient dense feeds have considerable potential as pig feeds, in particular for dry sows.

* Nutrient costs per hectare of land look promising compared to conventional feeds. It is unfortunate that set-aside land cannot be used for dry sows at low stocking rates (3 to 5 per hectare), as they would help to maintain land and provide a financial return.

* The main problems with feeding alternative feeds to pigs kept outdoors have been associated with:
a) matching nutrient intake to requirements;
b) handling and processing the feeds.

* Means of overcoming these problems should be developed; one approach could include complete diet feeding.

* The development of technology for low nutrient dense feeds could have other benefits, for example in improving health and welfare of the sow.

REFERENCES

ADAS (1984) *Fodder Beet*. Technical Bulletin TFS 346. Agronomy/Nutrition Chemistry, South East Region.

114

ADAS (1986) *Fodder Beet.* MAFF Technical Bulletin P591.

Carlisle, B. and Mitchell, W. (1984) *Maize Silage and Fodder Beet for Sows.* ADAS Nutrition Chemistry Technical Conference, 1984, 1-2.

Chambers, J. (1987) Feeding outdoor pigs; electronic sow feeders and other methods. *Proceedings of the Pig Veterinary Society*, **18**, 62-66.

Lee, P.A. and Close, W.H. (1987) Bulky feeds for pigs: a consideration of some non-nutritional aspects. *Livestock Production Science*, **16**, 395-405.

Livingstone, R.M. (1988) Sows prosper on bulky feeds. *Farmers Weekly*, 1 January 1988, 18.

Livingstone, R.M. and Fowler, V.R. (1984) Pig feeding in the future: back to nature? *Span*, 27 March 1984, 108-110.

Nix, J. (1989) *Farm Management Pocketbook* 19th Edition. Wye College, University of London.

Pond, W.G. (1987) Thoughts on fibre utilization in swine. *Journal of Animal Science*, **65**, 497-499.

Varel, V.H. (1987) Activity of fibre-degrading microorganisms in the pig large intestine. *Journal of Animal Science*, **65**, 488-496.

DISCUSSION

Although the potential for using low nutrient dense feeds in outdoor sow production appears considerable, the physical means of implementing the system are likely to be the main constraints. The continued improvement in electronic sow feeder technology, to provide supplementary feed, together with complete diet mixer trailer technology appears to offer the greatest hope for its implementation. Clearly a highly mechanised system would be needed to handle the bulky feeds proposed.

As the outdoor pig industry becomes more established, and whilst continuous economic pressure continues to be exerted on pig production in general, even greater emphasis will be placed on reducing the cost of pig feeds. The implications here are that low nutrient dense feeds will need to be considered with all other alternatives in the future, as the industry attempts to maintain its profitability and where feed comprises the major cost of production.

CHAPTER 8

FINISHING SYSTEMS FOR OUTDOOR PIG PRODUCTION

N M Beynon

SUMMARY

The Berkshire College of Agriculture profitably produces finished pigs from both indoor and outdoor sow units, with the latter being managed on a radial system. Indoor pigs are from Camborough White sows and outdoor pigs are from either Camborough Blue or Camborough 12 animals. The outdoor and indoor pig has similar basic requirements, such as food, water and maintenance of body temperature, but the weaned outdoor animal has experienced a totally different environment to that of the indoor pig. Behavioural differences resulting from being part of an outdoor unit need to be taken into account, in addition to genetic differences, when considering the management of weaned pigs. In particular, outdoor weaners tend to eat creep feed more readily post-weaning, to be healthier and to be more inquisitive, lively and destructive animals.

There are also several differences between finishing animals from the outdoor and indoor herds. At the Berkshire College progeny from the indoor herd and from the outdoor Camborough 12 sows are managed similarly, with a slightly restricted feed intake in the later stages of finishing; the carcases of the "improved" outdoor pigs are of similar quality to those from the indoor animals. Offspring of the Camborough Blue outdoor sows, however, are given a lower feeding level and taken to a lighter carcase weight to avoid carcase quality penalties. The carcase quality disadvantage previously incurred by outdoor pigs was often compensated by their lower cost of production. The improved outdoor hybrid has narrowed the carcase quality gap and given the outdoor breeder feeder an animal with the potential to procure a financial advantage over his indoor fellow. But with both types of finishing animals from outdoor stock, analysis of marketing records is extremely important to achieve optimum results.

INTRODUCTION

The Berkshire College of Agriculture has both an indoor and an outdoor sow unit, each containing about 100 sows. The indoor unit was restocked and the outdoor unit started, in 1985. Herd sizes are relatively small, due in part to their location at an agricultural college. Although management has to be adapted to accommodate teaching requirements, both units are run commercially and are profitable.

The outdoor unit is based on Camborough Blue sows, which have proved to be very useful animals, and Large White boars. The College was one of the first units to have Camborough 12 sows and, while HY boars have been tried, Large White boars are now used because of the unacceptably high level of stress deaths from the halothane positive HY sires. In the last 4 years the number of pigs reared has been between 21 and 23 per sow per year.

The College is now in the second year of a radial paddock system. The radial layout is based on paddocks with an outer perimeter track for vehicles delivering food, whilst a central circular passage provides gated access for the adult pigs (see Plate 8.1). The paddocks radiate outwards from the central passage and the electric fencing forms a pattern rather like the spokes on a wheel. This system allows easier movement of adult breeding pigs.

Plate 8.1 Outdoor boars and sows kept on the 'radial' paddock layout

The pig rotation is 8 years long and includes 4 years of arable crops, 2 years of grass, followed by 2 years of pigs, and then a return to arable. In the first year stubble was reseeded with grass, but there is now little cover remaining.

THE OUTDOOR WEANER

Essentially the outdoor pig has similar requirements to those of other pigs, but until weaning the outdoor animal is subject to a totally different environment compared to the indoor pig. The weather, climatic factors, structural accommodation and social factors, together with nutrition, husbandry and health pre-weaning all affect the behaviour and physical performance of the weaned animal.

Environmental factors contribute markedly to differences in behaviour between indoor and outdoor weaners, and these differences in behaviour should be taken into account, together with genetic differences, when considering finishing systems for outdoor pigs.

Compared to the indoor weaner, the outdoor pig:

* Is more inquisitive, active and, unfortunately, destructive;
* Is predisposed to take in water more readily at weaning;
* Is more easily weaned onto creep feed (with no previous feeds);
* Consumes more creep feed in the critical first two days post-weaning;
* Will transfer to second stage diets sooner;
* Tends to be heavier age-for-age at weaning;
* Mixes with less stress and aggression at weaning;
* Is healthier *per se*;
* Will grow more quickly post-weaning and will scour less;
* Does well in less sophisticated housing.

Young piglets reared outdoors probably consume up to 1.5 litres of milk per day, or 350 g milk dry matter, which would be equivalent to about 500 g of excellent quality creep feed. Good establishment of sucking by the piglet is important for high quality weaners, and this should occur within the first two days of a piglet's life.

Within a week the piglets are generally sufficiently active to scramble out of the ark, and they are then very curious to explore their environment. Outdoor piglets thus receive much more stimulation than their indoor counterparts. There may, however, be problems if the piglets leave the ark too early and are not sufficiently strong to withstand external conditions and to scramble back into shelter. In hot conditions sows may spend considerable periods of time at a wallow, if provided, and there is a risk that only those piglets strong enough to leave the ark and find the sow are adequately fed. This produces a dilemma, as it is important to discourage piglets from direct access to the wallow, in order to reduce the risk of drowning. A fender or piglet run needs to be used tactically in such circumstances (Plate 8.2).

Piglets learn to root outside very early in life and, although not fed creep feed, they rescue pieces of sow feed, often returning to their ark to chew a biscuit or nut (Plates 8.3 and 8.4). Piglets also generally learn to drink water. The outdoor piglet at this stage is inquisitive, active and learning.

Plate 8.2 A fender, or piglet run, designed to keep the piglets in, but allowing an easier re-entry via a step on each side if they do get out

Plate 8.3 Sows should be fed cobs, rolls or biscuits in winter or on wet soils

119

Plate 8.4 Outdoor piglets gain early experience of feeding, especially in summer (dry weather) when nuts can be fed instead of cobs or biscuits

At weaning the outdoor piglet, like the indoor animal, probably has a poorly developed digestive system and immune system, and relatively little control over body temperature. Essentially, therefore, at weaning the outdoor animal must be provided with its requirements as would an indoor piglet.

The Berkshire College pays great attention to ensuring an adequate water intake post-weaning, using a simple cube drinker to back up the drinking system (see Plates 8.5 and 8.6). This latter point is particularly important for outdoor weaners, which will not have seen a nipple-drinker before. Palasweet is also used for 4 days to encourage water intake from these cube drinkers (see Figure 8.1).

Providing attention to an adequate water intake has allowed the College to remove antibiotics from routine use in piglet feeds. Barber, Brooks and Carpenter (1988) have demonstrated that the water flow to drinkers has an important influence on daily feed intake (Table 8.1). At higher water delivery rates, growth rate increased and feed conversion efficiency improved.

The feed intake of piglets immediately post-weaning at the Berkshire College is nearly 50 percent higher for piglets reared outdoors compared to indoors. Estimated creep feed intake during the first 2 to 3 days post-weaning is 133 g per head per day for outdoor weaners and 90 g per head per day for indoor weaners, despite the latter usually having been fed creep feed pre-weaning (Herd Records, Berkshire College of Agriculture, 1988).

Plate 8.5 Weaner accommodation showing slatted dunging area, water nipples and a cube drinker—essential at weaning

Plate 8.6 Drinking from a nipple on a turbomat feeder

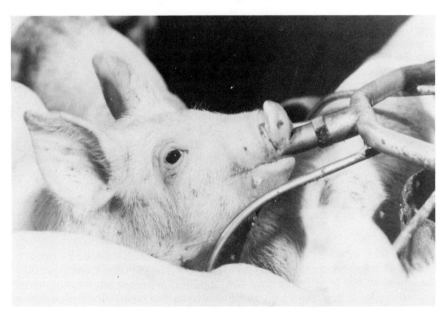

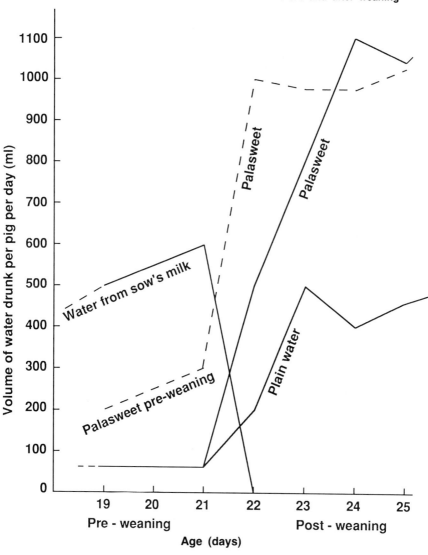

Figure 8.1   The effect of Palasweet on water intake before and after weaning

Source: Tate and Lyle (private communication 1988).

Feed intake is influenced by environmental factors, but at the Berkshire College environmental effects are standardised as both indoor and outdoor piglets are weaned into the same bungalows, which have underfloor heating. Table 8.2 illustrates the effects of floor type and environmental temperature on feed intake, which tends to be higher in slightly cooler conditions.

Table 8.1   Effects of water delivery rate on voluntary feed intake and water use (3 to 6 week old pigs)

| | Water delivery rate (ml/minute) | | | |
	175	350	450	700
Daily feed intake (g/pig)	303	323	341	347
Daily gain (g/pig)	210	235	250	247
Feed conversion efficiency	1.48	1.39	1.37	1.42
Daily voluntary water use (litre/pig)	0.78	1.04	1.32	1.63
Apparent time spent drinking (minutes/pig/day)	4.5	3	3	2.3

Source: After Barber, Brooks and Carpenter, cited by Paul Smith, Pigspec (1988).

Table 8.2   Effect of floor type and environmental temperature on the feed intake of weaned piglets

| | Environmental temperature (°C) | | |
Liveweight (kg)	Perforated metal floor	Straw bedded floor	Feed intake (g/day)
5	29	27	126*
6	29	27	157*
6	27	25	213**
7	29	26	142*
7	27	24	235**

*   Probable intake
**   Possible intake

Sources: After Bruce and Clark (1979) and Robertson, Clark and Bruce (1985).

Providing straw in effect decreased the lower critical temperature, resulting in similar feed intakes for the two systems, which were maintained at slightly different temperatures.

Figure 8.2 illustrates changes in the growth rate of piglets which occur pre and post-weaning. With both outdoor and indoor animals, immediately after weaning there is a marked depression in growth for several days, but growth rate rapidly increases again.

While environmental factors affect feed intake and growth rate, conversely the rate of growth of piglets influences heat production and the ability of the animals to withstand cold temperatures. Figure 8.3 shows the daily heat production of weaned piglets growing at different rates. When piglets are weaned into straw yards at the Berkshire College (see Plate 8.7), the outdoor piglets subsequently grow faster than those from the indoor herd, both

Figure 8.2 Daily growth rate of piglets from 4 days prior to 4 week weaning to 30 days post weaning

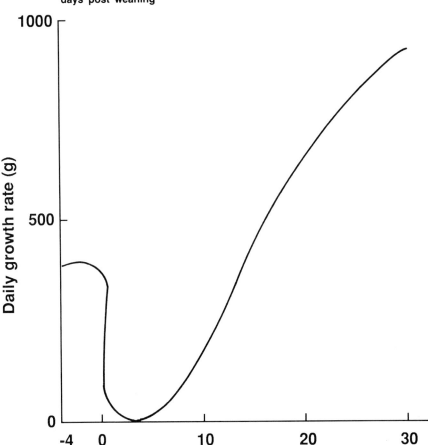

Source: Whittemore (1988).

because they are better adapted to the environment, and also as a result of the higher initial feed intake.

In general, weaners from the outdoor compared to the indoor unit tend to be heavier at weaning, are weaned more easily onto creep feed, grow rapidly with less scouring, mix with less stress and aggression, and transfer to second stage diets sooner. However, they are more inquisitive, destructive and ready to try to escape.

124

Figure 8.3 Influence of growth rate on the heat production of weaned piglets

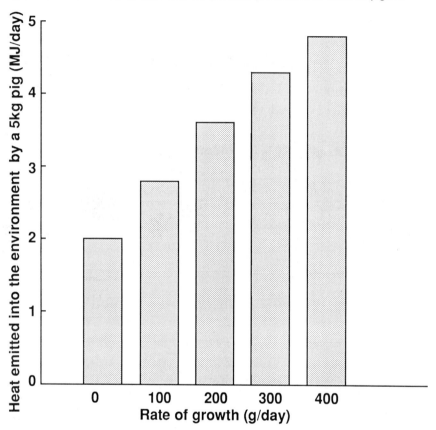

Source: Whittemore (1988).

THE OUTDOOR FINISHER

Two types of outdoor pigs are finished at the Berkshire College of Agriculture—the progeny of the Camborough Blue and the Camborough 12 sows. The former are typical of Blue pigs from outdoor units, whereas the latter represent the "improved" outdoor pig.

The feeding regime is similar for the Camborough 12 offspring as for the indoor pigs, and the two types of animals are often mixed together with no adverse effects. These finishers are given a maximum of 2.5 kg feed, which is virtually *ad libitum* intake, until they enter Trobridge pens. They are then floor fed and restricted to 2.5 kg feed. By contrast, the Camborough Blue

Plate 8.7 Simple straw kennel, open yard, ad-lib trough system for outdoor reared finishing pigs

Plate 8.8 A basic device for back fat measurements on the live animal

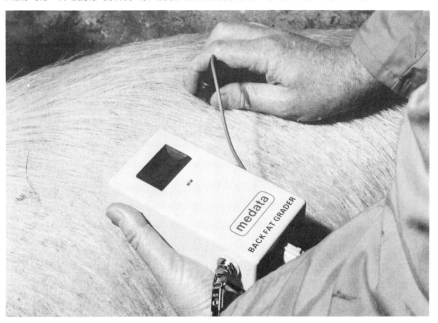

progeny are fed a maximum of 2.25 kg feed and, if accidentally mixed with indoor animals, adverse effects in the form of increased fat depths from the higher feeding level are readily observed (Plate 8.8).

Table 8.3 gives typical average backfat values for the various types of finished animals sent to J Sainsbury plc. Despite receiving a lower feeding level, the Camborough Blue offspring were slightly fatter, whereas backfat values of the progeny of the Camborough 12 sows were only marginally greater than those of the indoor pigs fed the same quantity of feed. All animals were sold at 70 kg carcase weight.

Table 8.3   Typical backfat values for finished pigs of 70 kg weight from the indoor and outdoor units

Unit	Type of pig	Entires	Gilts
Indoor	Camborough White	22.5 mm	24.5 mm
Outdoor	Camborough Blue	24.5 mm	27.75 mm*
Outdoor	Camborough 12 (Duroc cross Landrace)	23.5 mm	25.5 mm

* Fed maximum of 2.25 kg/day compared to 2.5 kg for other pigs.

Source: Berkshire College of Agriculture Herd Data (1988).

Using 1988 pig prices, the 'optimum' average carcase weight was about 72 kg for the entire indoor pigs and 66 kg for Camborough Blue gilts. However, at a higher price per kg carcase, the optimum weights would increase to 77 kg and 70 kg respectively, based on computer analysis.

Figure 8.4 shows the relationship between backfat thickness and carcase weight for entire males and gilts from the outdoor and indoor units at the Berkshire College of Agriculture. Although at any given carcase weight outdoor animals were fatter than the same sex pig from the indoor unit, finished entires from the outdoor sows (Camborough Blue) were slightly leaner than gilts from the indoor unit. Over the range of carcase weights, from 63 to 73 kg, all $P_1 + P_3$ measurements were satisfactory.

Table 8.4 illustrates the detailed analysis provided to herds marketing via the CAMBAC producer group. A slap mark is used to identify pig type or special treatment and an analysis of both male and female pigs is provided, including a financial comparison.

Figure 8.5 is a computer generated graph which illustrates the financial penalty of taking the average back fat of pigs of the same weight from 21 to 29 mm $P_1 + P_3$. In this example it costs around £2 per pig if back fat thickness deteriorates by this amount. These tabulated figures are an example of the type of analysis produced by the computer based pig marketing model developed by Beynon (1985).

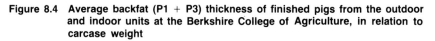

Figure 8.4 Average backfat (P1 + P3) thickness of finished pigs from the outdoor and indoor units at the Berkshire College of Agriculture, in relation to carcase weight

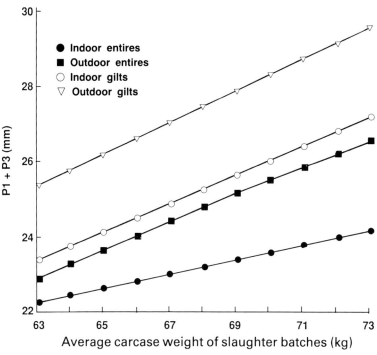

● Indoor entires
■ Outdoor entires
○ Indoor gilts
▽ Outdoor gilts

P1 + P3 (mm)

Average carcase weight of slaughter batches (kg)

Source: Beynon (1988).

The progeny of both the indoor and outdoor pig herds (Camborough White and Camborough Blue sows, respectively) can be finished profitably, and hopefully offspring from Duroc crosses (e.g. Camborough 12) will increase profit margins compared to the Camborough Blue. However, analysis of marketing records is important so that the sale of over-fat animals and the ensuing financial penalties are avoided.

Records from CAMBAC provide information on the average carcase weight and backfat thickness of groups of pigs, together with the average price received and the best price possible. Further data are provided to indicate why the optimum slaughter price was not achieved (see Figure 8.5).

The Berkshire College has carried out trials recently with a β-agonist, which reduced backfat thickness by 1 to 2 mm. This therefore appears to be a potentially useful tool, although political decisions may prevent its application.

Table 8.4 Cambac analysis summary

Slap Mark	Number of		Average P_1, P_3 (mm)		Average carcase weight (kg)			Price received (p/kg carcase)			Whole condemned carcases		Condemned parts	
	Gilts	Boars	Gilt	Boar	Gilt	Boar	Overall	Best	Average	Difference	Number	Weight (kg)	Number	Weight (kg)
RO361	80	63	27.0	25.6	71.7	72.2	71.5	99.60	96.49	3.11	0	0.0	8	27.0
RO241	3	3	21.7	20.7	69.7	68.7	69.0	101.60	98.84	2.76	0	0.0	0	0.0
RO124	0	2	0.0	23.5	0.0	68.2	68.0	101.60	98.38	3.22	0	0.0	0	0.0
RO123	7	15	21.6	22.7	68.0	68.9	68.5	101.60	98.96	2.64	0	0.0	0	0.0

Source: Cambac Pig Sales, Checkenden, Reading.

Figure 8.5 Optimum slaughter weight analysis

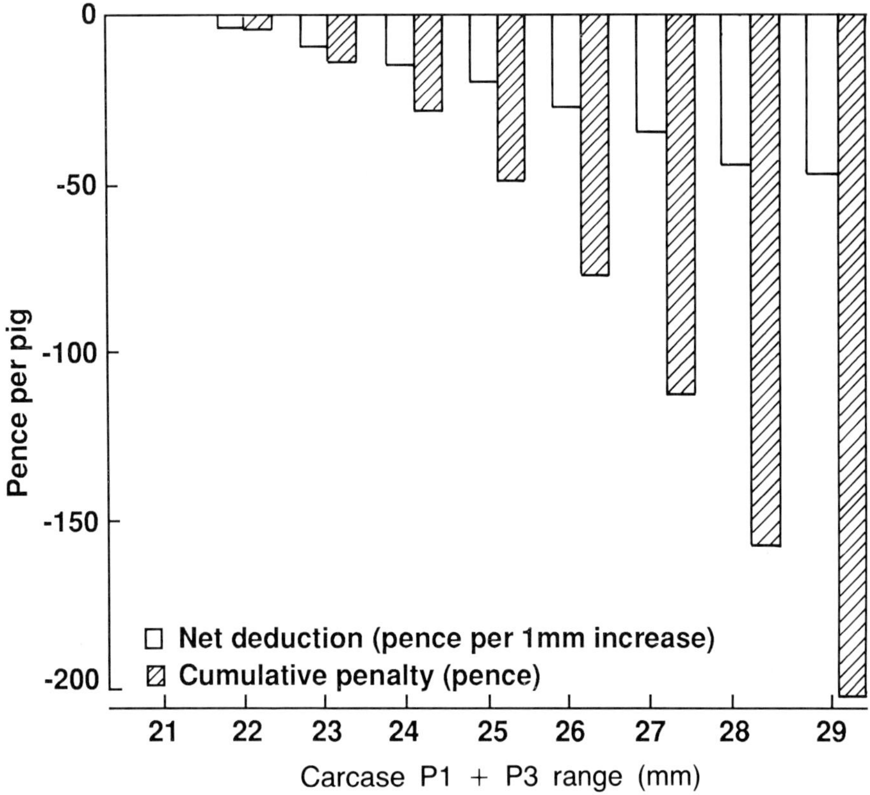

Source: Beynon (unpublished data).

CONCLUSIONS

The actual back fat figures and the computer analysis both indicate that the probe depths on outdoor blue gilts are significantly higher. This results in a carcase grading penalty of around £1.60 per pig between outdoor Camborough Blue offspring gilts and the White hybrid indoor entire male pigs (see Table 8.3 and Figure 8.5). This carcase quality disadvantage is often compensated by the lower cost of producing and possibly housing the outdoor weaner. The 'improved' outdoor hybrid has narrowed the carcase quality gap and given the outdoor breeder feeder an animal with the potential to procure a financial advantage over his indoor fellow. Computer analysis of the carcase grading data can now easily provide information indicating the best weight to market into selected carcase grading schemes. Optimum weights and back fats will vary according to pig type, contract specification and pricing. The outdoor weaner and finisher require just as much management input as their indoor cousin, and attention to detail will help to achieve optimum results.

It is important to note that the points made in this paper relate to a unique situation in the Berkshire College herds, where outdoor weaners are managed in the same buildings and on the same feed types as the indoor pigs. Marketing through the same outlet also provides a useful comparison of carcase characteristics.

REFERENCES

Beynon, N.M. (1985) *Evaluation of Graded Market Outlets for Slaughter Pigs Using Micro-Computer Based Statistical Models.* Thesis submission for the Insignia Award in Technology, London 1985.

Beynon, N.M. (1988) Contribution to *A Seminar on Pig Marketing*, organised by J. Sainsbury plc, Cambridge, March 1988.

Bruce, J.M. and Clark, J.J. (1979) Models of heat production and critical temperature for growing pigs. *Animal Production*, **28**, 353-369.

Robertson, A.M., Clark, J.J. and Bruce, J.M. (1985) Observed energy intake of weaned piglets and its effect on temperature requirements. *Animal Production*, **40**, 475-479.

Whittemore, C. (1988) Pig Farming/BOCM Silcock 'Extra', March 1988.

DISCUSSION

The restriction of Camborough Blue offspring to 2.25 kg feed leads to an increase of 7 to 9 days in time to slaughter, and this is generally worthwhile as back fat levels are reduced. Food conversion ratios are not adversely affected. The Berkshire College also finds that the outdoor pig copes better than its indoor counterpart when finishing is carried out in older buildings, with slightly inferior environmental conditions, therefore building costs per pig produced per pig place are reduced.

Although offspring from the Camborough Blue can be managed so that an over-fat carcase does not arise, in practice the carcase tends to look plainer than that from improved breeding stock. The Berkshire College has, however, experienced no adverse reactions by carcase purchasers in this respect.

The importance of analysis of market data must be emphasised, particularly differences between groups of pigs and the range of carcase weights and backfat values recorded. Analysis allows the manipulation of subsequent finished pigs through adjusting diet and feed scale, to produce the required product and to achieve good margins.

CHAPTER 9

BREEDING FOR OUTDOOR PIG PRODUCTION

M Bichard

SUMMARY

The requirements for the outdoor herd reflect both the physical conditions in which the sows and boars live and the needs of the finisher and final buyer. Inevitably a compromise is necessary, but the nature of the solution changes with time.

Sows must be hardy for year-round exposure to the environment, prolific to ensure high output and have the character which leads to ease of management. This latter involves docility for moving and handling, and good maternal instincts to make a nest, defend it against other sows and care for the young piglets. Boars need to be strong enough to withstand rough conditions underfoot, to have sufficient libido to work well in all weathers and to be active without being too aggressive. The progeny should be vital and active from birth, capable of rapid and efficient growth from weaning to slaughter, and produce a desirable carcase in terms of fat level, lean content and quality.

The balance between these requirements involves a trade-off. In the past the outdoor herd owner was prepared to accept a relatively poor carcase in return for the ability to survive without too much management, and to have animals with high feed intakes. As discounts for poor pigs become greater, a smaller proportion are sold to swill feeders and more are finished by the breeders, the balance must tip in favour of a good carcase quality. The breeder is concerned to provide the optimum balance of genetic potential, and to do this in the right 'package'. Aspects of this 'package' include the regular supply of both sexes, or of boars only; the age and weight of boars and gilts on delivery; and the health and condition (readiness to live and work outdoors in groups) of the replacement animals.

INTRODUCTION

Outdoor pig production systems are less well defined than for the indoor situation, which leads to several problems in providing breeding stock for this sector. A crucial question is the type of animal wanted for the outdoor situation. In general, breeding companies have expertise in selecting genetic lines and other breeding techniques, but information is needed from outdoor producers themselves on the relative importance of different characteristics. Breeding is a long term activity and the current breeding programme determines the type of animals that will be kept outdoors 5 years hence.

REQUIREMENTS FOR OUTDOOR SOWS AND BOARS

Factors to consider for outdoor breeding stock include:

* *Normal genetic requirements for breeding stock*

  a) Prolificacy in sows
  b) Libido in boars
  c) Fast, economic growth and good carcase quality in the slaughter generation

* *Additional genetic needs for outdoor units*

  a) Hardiness in sows and boars
  b) Character and intelligence in sows, for example to aid the movement of animals around the unit
  c) Ease of farrowing in sows
  d) Ability of sows to provide maternal care

* *Physical features of breeding stock supplied*

  a) Appropriate age and weight
  b) Healthy
  c) Good body condition and ready for outdoor life, for example by boars already being kept in groups

CONFLICTS BETWEEN BREEDING OBJECTIVES

There must, of necessity, be compromises between some of these requirements, particularly in terms of carcase quality. Most prolific lines do not have the best carcases, and neither do hardy sows and boars. Animals with a docile character also tend to have poorer carcases. Whilst sows which have a good appetite are able to consume large amounts of feed during lactation, and thus lose relatively little body tissue, if their progeny also have high *ad libitum* feed intakes over-fat carcases may result.

In general, the main conflicts between characteristics which are considered in breeding programmes for outdoor stock are:

Favourable characteristic   *Associated disadvantages*

Rapid growth rate   Fatter carcases.

Favourable characteristic (cont)	*Associated disadvantages (cont)*
Heavy muscling	Lower litter size. Poorer physical soundness.
	Increased susceptibility to stress deaths. Drip loss from carcases.
High lean content of carcase	Reduction in flavour and juiciness of meat, perceived by taste and consumer panels.
Coloured skin to protect against the outdoor environment, particularly sun-burn	Considered by carcase purchasers to produce a fatter carcase.
Good sow behaviour and intelligent animals	Coloured sows (or grandparents).
High lean growth rate	Larger mature size of boars and sows.
Self-limiting appetite in finishing pigs	Inadequate appetite in sows, leading to excessive body tissue loss in lactation and subsequent re-breeding problems.

These are not genetic correlations within a line, but rather observations across a range of breeds and lines.

Conflicts in breeding objectives have changed with time. The following trends in pig production during recent years have important implications for outdoor pig producers and the type of stock they require. Thus the best comprise now between different characteristics is not the same as it was a few years ago.

Reduced demand from swill feeders for heavy, store pigs

These animals were required to have a large appetites, and the offspring from outdoor herds ideally suited the needs of swill feeders.

More animals finished by breeders

There has been a general trend towards larger pig units, with both the breeding and rearing of young animals taking place within the same enterprise, to improve labour utilisation and business efficiency. Thus, many breeders have become more aware of the need for progeny with good carcase characteristics.

Improved carcases from indoor herds

Each year the fat cover of pigs in the UK reduces by 0.5 mm. Consequently the standard by which outdoor pigs are judged is constantly changing. Carcases which were acceptable 11 years ago are now no longer so.

Larger penalties for poor carcases

Carcase prices increasingly reflect carcase quality, and financial penalties for over-fat carcases have increased in relative terms.

Price/cost squeeze within the pig industry

Margins for finishing animals now fluctuate more widely than previously, but generally follow a downward trend. Higher output and efficiency are expected from outdoor herds to combat this price/cost squeeze.

Higher management levels

In general, herd management has improved in response to the need for higher levels of output.

Earlier weaning

Most piglets are now weaned at 3 weeks of age rather than the traditional 8 weeks, to increase the potential number of litters per sow per year and sow output.

Improved health status of pig herds

It is now appreciated that the health status of animals kept outdoors can be maintained at a high level, although the relative importance of different diseases and disorders varies between outdoor and indoor systems.

PERFORMANCE LEVELS IN OUTDOOR HERDS

Performance levels achieved under the MLC Pigplan scheme are shown in Table 9.1 for herds farrowing outdoors, and in Table 9.2 for breeding herds kept under intensive conditions. The figures are for the 12 months ending June 1988. With PIC stock, 14600 sows farrowing outdoors averaged 2.25 litters per sow per year, with 21.5 pigs reared per sow per year and an average of 9.57 pigs reared per litter. These results contrast with figures for PIC indoor herds of 2.30 litters per sow per year, with an average of 9.73 pigs reared per litter and 22.4 pigs reared per sow per year. Sows in PIC outdoor units were actually more productive than the non-PIC stock in indoor herds. Thus, outdoor units can no longer be regarded as low input/low output systems.

Table 9.1   Breeding herd results for sows farrowing outdoors (12 months to June 1988)

	PIC stock	Non-PIC stock
Total number of sows	14,595	3,756
Litters/sow/year	2.25	2.21
Pigs reared/sow/year	21.5	19.7
Average litter size (alive and dead)	11.28	10.60
Average number of pigs reared/litter	9.57	8.94

Source: Analyses of Pigplan results by MLC, June 1988.

Table 9.2   Breeding herd results for sows kept indoors (12 months to June 1988)

	PIC herds	PIC top third of herds	Non-PIC herds
Total number of sows	40,755	12,480	103,602
Replacements (%)	39.1	38.9	40.0
Sow mortality (%)	3.2	2.8	4.0
Farrowing (%)	88.2	88.8	85.4
Litters/sow/year	2.30	2.41	2.26
Pigs reared/sow/year	22.4	24.6	21.3
Average litter size (alive and dead)	11.72	12.21	11.41
Average number of pigs reared/litter	9.73	10.21	9.41

Source: Analyses of Pigplan results by MLC, June 1988.

The performance results for rearing herds given in Table 9.3 are for animals entering the rearing period at about 6 kg liveweight and passing to the finishing stage at 32 to 33 kg liveweight. Figures are very comparable for indoor and outdoor herds, and the mortality rate, at least in PIC herds, appears to be lower for outdoor compared with indoor systems. While a sale-price advantage might be expected for alert, healthy animals reared outdoors, this is counteracted by buyers considering that it is more difficult to produce a good carcase from outdoor stock.

Table 9.4 gives carcase data on the progeny of a variety of PIC outdoor gilts with PIC terminal sires. It is possible to achieve good carcase gradings from pigs from outdoor units, fed on a restricted regime. However, there is a tendancy for finishing pigs to be taken to lighter weights and the objective now is to provide animals which allow outdoor producers to be as flexible in marketing as indoor producers.

Results from Farm BC illustrate the effects of changing from Camborough Blue to Camborough 12 sows on the carcase quality of the progeny. The reproductive performance of the hybrid HY boars is very satisfactory. Naturally, that of new types of sows needs to be monitored, as there would

Table 9.3 Rearing herd results for pigs born in outdoor and indoor systems

	Farrowing outdoors		Farrowing indoors	
	PIC stock	Other stock	PIC stock	Other stock
Number of herds	18	8	89	243
Average number of pigs	1267	400	1057	593
Average start weight (kg)	6.1	5.7	6.1	6.1
Average final weight (kg)	32.1	31.2	33.5	29.9
Average daily gain (g)	434	424	445	408
Feed efficiency ratio	1.85	1.86	1.77	1.74
Mortality (%)	1.9	3.6	2.2	2.7
Average sale weight (kg)	32.8	31.8	31.3	32.4
Average sale value (£)	29.88	28.30	29.05	28.94

Source: Analyses of Pigplan results by MLC, June 1988.

Table 9.4 Carcase results from progeny of PIC outdoor gilts by PIC terminal sires

Progeny from		Average number of pigs	Average carcase wt (kg)	Average backfat P2 equivalent (mm)
Farm P	Camborough Blue	123	54.8	14.5
	Camborough 12	167	64.1	12.4
Farm BC	Camborough Blue[1]	200	67.0	13.1
(HY boars)	Camborough 12[2]	67	67.0	12.2
Farm L	Mixed	9,626	54.0	11.0
		2,830[3]	46.9	10.3
		5,028[3]	54.6	11.1
		1,723[3]	63.8	11.8
Farm C	Mixed	10,108	49.5	11.2

Notes: [1]Fed maximum 2.25 kg/day
       [2]Fed maximum 2.5 kg/day
       [3]Total pigs sub-divided into groups according to weight

Source: PIC customer data.

be little benefit from the progeny of Camborough 12 sows having better carcases than the offspring of Camborough Blue sows if, at the same time, the number of piglets reared per sow was reduced.

Performance data for Camborough 12 sows are given in Table 9.5, and compared with results from Camborough Blue animals. The data refer to about 3000 litters on a number of farms and show small differences between the Camborough 12 and Camborough Blue.

Table 9.5 Field performance averaged over first four litters of Camborough 12 compared with Camborough Blue sows

Litter size	Camborough 12	Advantage over Camborough Blue
Total born	11.46 (0.12)	0.38 B
Born alive	10.76 (0.12)	0.24 B
Born dead	0.70 (0.06)	0.14 W
Weaned (ignoring fosters)	9.49	0.02 W

Notes: Standard error given in brackets

B = better and W = poorer performance than Camborough Blue

Source: PIC field trials involving over 3,000 litters.

Carcase quality may be improved by a number of different methods:

* Continuous testing and selection in basic stock lines. Backfat thickness has been reduced in the Saddleback in recent years, and it would seem that by managing animals better and by more precise matching of feed to requirements, outdoor sows may not need to carry such large fat reserves as previously thought necessary.

* Introduction of female stock lines with
(a) more lean tissue
(b) higher killing out percentage
(c) better conformation.

* Use of improved sire lines with
(a) more lean
(b) less fat
(c) less bone.

Breeding companies are looking for stock lines which contain more lean tissue yet have a higher killing out percentage. Although the Saddleback carcase has a relatively high fat content, the killing out percentage is relatively low, and this is due at least in part to a poor muscle/bone ratio.

The most important carcase features are currently considered to be the contents of lean tissue, fat and bone. A terminal sire contributes 50 percent to these features, thus the use of an improved sire line can lead to marked carcase improvements.

However, when meat quality is assessed by trained taste panels and consumer panels it appears that the eating quality of slightly fatter carcases is better. At present sub-cutaneous fat thickness is the only criterion used to assess carcase quality, and there may in future be a need for additional criteria based on results from taste and consumer panels.

PRESENTATION OF BREEDING STOCK

Outdoor systems are very resilient and their low input characteristic is likely to continue. Breeding companies must realise this and supply stock suitable for simple management. A number of options exist, however, to meet the needs of outdoor producers.

The way in which to present genetically improved breeding stock to the purchaser needs to be considered. Outdoor herds generally use specialised dam and sire lines to provide different characteristics. There are always two options for outdoor producers: either the unit can maintain small herds of specialised lines, with crossbred parent gilts being bred in the unit, or all replacement gilts and boars may be bought from specialised multiplier herds. Where herds choose to purchase all replacements then animals may be introduced at a variety of ages, for example weaner gilts or maiden gilts.

Breeding companies now accept that boars must be sold in working groups which are used to each other and already acclimatised to the outdoor situation.

DISCUSSION

At present there is probably too much emphasis on the need to provide animals to live in a cold environment outdoors. Heat is often a problem in summer, and the animal's ability to withstand high temperatures is at least as important as resilience to cold.

Producers need to convey their requirements for stock characteristics clearly to the breeding companies. The Pig Improvement Company relies on this type of communication, as it believes that more knowledge is gained from widespread contact with outdoor producers than by the company maintaining its own outdoor units.

Two big differences between animals in outdoor and indoor herds are backfat thickness and appetite. Although not a directly selected trait, the appetite of pigs in indoor herds has reduced over the years as a consequence of reducing backfat. With outdoor sows, a good appetite is needed to compensate for lower fat reserves in animals selected for improved carcases. The relationship between backfat levels and susceptibility of animals to different stresses is largely unknown.

There is concern amongst some that by reducing backfat levels too much, meat quality may be impaired to the extent that consumers reduce their purchases. Different subcutaneous fat depots have a relatively high genetic correlation, but there is evidence that between-breed differences exist in other fat stores; some lines partition a higher proportion of their fat into intramuscular stores at a given subcutaneous fat level. However, the relationship between intramuscular fat and eating quality is not well established and it is likely that other factors in the lean tissue itself also have an influence on eating quality.

Great Outdoors

The new Camborough 12

CHAPTER 10

OUTDOOR PIG PRODUCTION, ANIMAL WELFARE AND FUTURE TRENDS

A J F Webster

SUMMARY

The term "welfare" may be used to refer either to the perception of the animal's environment by the welfare lobby or to the perception of the environment by the animal itself. The distinction between the two interpretations is important, and this paper concentrates primarily on the latter, namely the animal's perception of its environment. Implicit in the Farm Animal Welfare Council Codes is the fundamental theme of the five freedoms: freedom from malnutrition, freedom from thermal and physical discomfort, freedom from injury and disease, freedom from fear and stress, and freedom to express most normal, socially acceptable patterns of behaviour. These present a practical check list by which to assess welfare within any husbandry system, and which can form the basis of a decision on the best compromise between sow comfort, health and productive efficiency. With outdoor sows parasitism, abnormal behaviour, injury risks and hygiene can be controlled with good management and stockmanship, but variations in environmental conditions and thus thermal comfort are areas of concern. Within the UK, air temperature per se is of relatively little importance, but wind speed and precipitation have major effects in relation to cold stress, while direct sunshine is the major contributor to heat stress. In the practical outdoor situation, hot sunny conditions are probably of more significance than cold weather.

Any system of animal husbandry can create welfare problems. Outdoor pig production may appear to the consumer to be a healthier and happier process, but may not necessarily be seen as such by the sow. The only way to ensure improved welfare standards for farm animals within any husbandry system will be to impose a set of clearly defined rules, properly enforced throughout the European Community (e.g. by licensing), that will free farmers from the crudest excesses of the free market and allow them to practise husbandry standards of which they can be rightly proud. As many of the characteristics which are favourable for sows kept outdoors are detrimental to the slaughter generation, the best option for outdoor pig production in the future may be to produce breeding animals which are well-suited to the outdoor environment, and to manipulate lean tissue growth and nutrient deposition in the progeny by non-genetic means, particularly through the use of hormones.

INTRODUCTION

The term "welfare" is often used with two quite distinct meanings, the first referring to the perception of the animal's environment by the welfare lobby and the second relating to the perception of the environment by the animal itself. These two interpretations of "welfare" bear little relationship to one another.

According to the Welfare Codes, the animal's welfare needs are met by its environment if the animal can adapt to that environment without suffering, at the least, or, rather better, if it expresses some form of positive well-being in that environment. This reaction by the animal probably differs from the human perception of how a sow sees its environment, and these differences are very important for pig producers for two reasons. Firstly, the belief of the welfare lobby may differ very much from the perception of the sow and create forms of legislation designed to meet a consumer need which may actually be detrimental to pigs themselves. Secondly, the active welfare lobby, which is the group which affects pig producers financially, is quite small despite its volubility. Some consumers refuse to buy animal products, while others demonstrate their concern about animal welfare by buying organic products or welfare-improved meat products; supermarket evidence together with consumer polls conducted by organisations such as the Meat and Livestock Commission suggest that the latter probably comprise between 6 and 8 percent of consumers who purchase animal products.

The welfare lobby is not large and, unless legislation is enforced, its influence on pig production is relatively small. A new market may be created for producers to provide meat to satisfy those people who wish to eat meat with a clear conscience. However, this market is very small and, while meeting the perceived needs of the consumer is a good marketing strategy, it does not help the 90 percent or more of farm animals which may be in farming systems which have welfare disadvantages. In the case of the sow, the number of alternative husbandry systems that can work successfully is considerable, and this latter problem is of relatively little significance. By contrast, at present there is no alternative which competes economically with battery cages for the laying hen and thus the move to free-range eggs by supermarkets is having a marginal effect on the welfare of most birds.

WELFARE AS PERCEIVED BY THE ANIMAL

Attempts have been made by the Farm Animal Welfare Council to define welfare-related terms and these are published in the Farm Animal Welfare Codes, copies of which all pig producers should have received. Implicit in these codes is a fundamental theme, the five freedoms. Although these are ideals, I believe that they are more than just pious hopes and that they can be used as a basis for a very logical evaluation of different husbandry systems.

The criteria expressed in the five freedoms are:

* Freedom from malnutrition.
* Freedom from thermal and physical discomfort.
* Freedom from injury and disease.
* Freedom from fear and stress.
* Freedom to express most normal, socially acceptable patterns of behaviour.

It is clearly inadequate to state that the physical comfort of sows in individual stalls with no bedding is poor. The concept of the five freedoms can be used to set up a matrix by which any husbandry system can be assessed and a "least worst" solution developed, which meets minimal standards for the welfare of animals in particular systems. Table 10.1 illustrates how the advantages and disadvantages of different systems can be compared; for example, there may be a conflict between allowing more normal behaviour but having a greater risk of injury.

Table 10.1 Effect of housing system on the welfare of dry sows

	Paddocks and arks	Individual stalls (no bedding)	Covered straw yards
Thermal comfort	Very variable	Fair to poor	Good
Physical comfort	Variable	Bad	Good
Injury	Slight	Feet, 'bed sores'	Fighting
Hygiene	Fair to poor	Usually good	Fair to poor
Disease	Some parasitism-control difficult	Usually good	Parasitism-control easy
Abnormal behaviour	Slight	Severe	Slight

The logic behind the approach of the Farm Animal Welfare Council is that it attempts to be relatively comprehensive, thus avoiding considering welfare only in terms of productivity criteria, which may be the view of some in livestock production, or only in terms of behaviour, which is the view of the majority of the welfare lobby.

With outdoor sows, parasitism can be controlled, and abnormal behaviour and injury risks can be reduced by using the appropriate genotype, correct management and good stockmanship; hygiene is variable but can be controlled. A factor, however, which is repeatedly of concern is thermal comfort.

BODY TEMPERATURE CONTROL

Cold and heat are frequently equated with temperature, and critical temperatures are often equated with the zone of thermal comfort. This is,

144

however, an inadequate description of heat exchanges between a sow and her environment, especially outdoors.

The sow produces heat, and the more she eats the more heat she produces. A bigger problem in practice is heat rather than cold stress, which leads to a paradox in animal production systems. Livestock which are kept most intensively, in terms of high stocking densities in relatively high temperature environments, namely pigs and poultry, are the animals that are most susceptible to heat stress because they have a very limited capacity to lose heat by evaporation. Neither sweat to any extent, nor are able to lose heat by thermal panting.

Figure 10.1 Patterns of heat exchange in farm animals

H_p = heat production; H_1 = total heat loss; H_e = heat loss by evaporation; H_n = sum of sensible heat loss by conduction, convection and radiation.

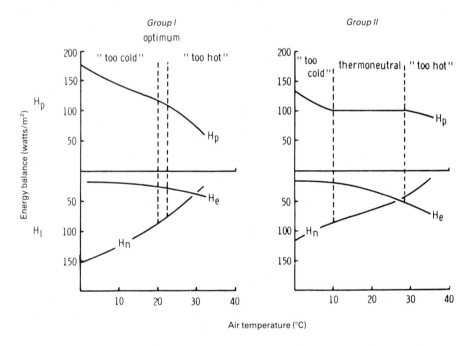

Group I animals maintain homeothermy primarily by regulating heat production (H_p), e.g. pigs and chickens. The vertical lines indicate the narrow zone of air temperature wherein food conversion efficiency is optimal.

Group II animals maintain homeothermy primarily by regulating evaporative heat loss (H_e), e.g. horses and ruminants.

Source: Webster (1983).

In her natural environment a sow regulates metabolic heat production, in order to keep body temperature up to the desired level, by regulating food intake. Man and cattle regulate evaporative heat loss to keep body temperature down to the required level.

This thermal regulation is normal and non-stressful, and takes place with no adverse effects even in severe cold conditions, as perceived by man. Pigs or farmyard hens wandering around at 10 to 12°C on a still day are not cold, although they are below their critical temperature. They are not cold in a welfare sense nor in a physioiogical distress sense, but they are using slightly more food energy in order to maintain body temperature. This is achieved without shivering.

The thermo-neutral zone for pigs and poultry is a very narrow zone which lies immediately below the point at which these animals are too hot. It is the zone of maximum food conversion efficiency, but it is not synonymous with the zone of thermal comfort. The pig can be perfectly comfortable well below the critical temperature, so long as adequate food is available. The design of foods appropriate to the comfort and satisfaction of the outdoor sow is a major subject for development.

Within the UK, cold temperatures *per se* are of relatively little importance. Wind speed has a greater effect on heat loss; for example, a 5 mph wind speed is equivalent to a 14 to 15°C drop in air temperature. In addition, while lying down the sow may lose 40 to 50 percent of heat by conduction directly to the floor surface if its conductivity is very high. One of the main reasons why concrete sow stalls are so bad for the welfare of the sow is that more heat is lost by conduction through the floor than by convection to the air. Thus the insulating power of the floor surface and the extent of wind and rain are far more important than the air temperature.

In a properly designed outdoor unit with well-bedded arks, the sow can control her environment to avoid the main cold stresses of wind and precipitation from above, while heat losses to the ground are minimal.

The problem of heat stress for the pig is that it has no effective physiological mechanism for regulating heat loss by evaporation. The sow can do little about heat stress in the short term except wallow; in the longer term, she will reduce her metabolic rate by reducing food intake. In the UK, heat stress for the sow is not usually a problem unless she is in direct sunlight, as the number of very hot days without direct sunlight is small. Such conditions are usually humid, but this is of relatively little importance as the sow loses so little heat by evaporation.

Pigs may become hot either

a) because the air temperature is high, or
b) due to the radiant heat load from direct sunlight.

Even in a hot environment air temperature *per se* is relatively unimportant, and effective wallows provide an effective remedy. Shade from direct sunlight is important to reduce heat stress. In some countries with cold winters and hot summers, the problems of mid-summer infertility seem to be little worse than in the South of England and Northern Europe, which suggests that, while there may be a genetic effect, micro-environments have been created which allow the sow to:

a) Get out of the sun
b) Increase evaporative heat loss by lying in mud, which is more efficient than water alone, as the duration over which water evaporates from the surface is prolonged. There are, however, problems with wallows if the mud is too deep.

If an animal gets soaked in winter conditions there is a large increase, some 3 to 4-fold, in heat loss. Therefore, in winter severe problems may arise with muddy, wet animals. Low air temperatures *per se* in the UK are likely to have relatively minor effects as pigs, like cattle, have a very good capacity to adapt to cold. It is other factors, such as wind and rain, which are more likely to cause cold stress.

Body condition

Sows which are in better body condition, with a greater fat cover, probably eat more food, produce more heat and are thus more tolerant to cold conditions. In addition, they possess more thermal insulation against the cold and also, when exposed to cold, the tissue metabolized contains a greater proportion of fat and a lower proportion of lean. A good fat covering is therefore cost-effective, as a sow which is thin and exposed to cold quickly begins to break down lean body tissue and rapidly to lose body condition. To maintain body condition in cold conditions an animal must be relatively fat.

In relation to heat stress, body fat has a marginally beneficial effect against the direct effects of solar radiation, but generally sows with more body fat have a higher level of heat production and a lower capacity for heat loss by convection through the skin.

Insulation

Baby piglets can cope well even when temperatures outside their shelter are low, and the question must be asked whether insulation of ark roofs is worthwhile. When piglets lie huddled inside their shelter, a major contributing factor to heat loss is conduction through the floor. Adequate straw bedding, and in some instances deep litter, is important to minimise this loss of heat; in addition, burying into straw reduces convective losses to the surrounding air.

Any type of sound roof prevents rain reaching animals inside the shelter, and also has a large effect on reducing the chilling effect of frost. On a clear, frosty night the external radiant temperature may be 30 to 40°C lower than the temperature inside the ark, whereas insulation of the roof may only increase the inside temperature by a further 3 to 4°C. Under these circumstances, insulation of the roof of arks may not be cost-effective.

CURRENT AND FUTURE ISSUES IN SOW WELFARE

The threat to animal production by the welfare lobbies is that of legislation. The Minister of Agriculture has responded to the Farm Animal Welfare Council by stating his intention to work on a ban to prevent the further construction of confinement stalls for dry sows. The position of dry sow stalls has until recently been strengthened by the fact that it is an economically competitive system, but now other systems, such as outdoor units, offer financially viable alternatives. Once the pig industry shows workable alternatives, the case for legislation against sow stalls strengthens.

Another possible change in response to welfare issues is the licensing of livestock premises and regular inspections.

The Farm Animal Welfare Council suggests a minimum of three-week (probably 19 to 23 day) weaning, but at present little attention is being paid to the piglets produced from welfare-oriented sow units.

The family group pen and other welfare-motivated systems of pig production are not realistic in their present form. There is, however, an interesting aspect of the family group which does not appear to have been taken up; over a number of years there has been a very successful return to oestrus during lactation, which holds potential for other, more commercial systems. Nutrition, particularly in relation to increasing heat production and providing digestible fibre, is also a part of sow welfare.

FUTURE TRENDS IN OUTDOOR PIG PRODUCTION

The outdoor system of pig production is a natural system and the potential to manipulate nutrition or the environment is limited. Thus full advantage must be made of the scope which exists for genetic manipulation of the pig itself. A major problem with pig breeding is that favourable traits in the slaughter and breeding generations are often conflicting, and pigs for indoor units have been selected primarily for favourable traits in the slaughter generation, with the environment being modified to accommodate the slaughter generation type sow.

A different situation occurs with outdoor systems. Improvements in charac-teristics such as prolificacy and temperament are likely to arise from continued conventional breeding and selection programmes, although there is a possibility of increasing prolificacy in the sow through genetic engineering. If piglets are bred with a greater dependency on traits more suited to animals kept outdoors, deficiencies in product quality and production efficiency are likely to arise in the slaughter generation.

Rather than manipulating the slaughter generation entirely by manipulating the characteristics of the boar, it is possible to manipulate the slaughter generation by the use of hormones at the appropriate time. This means that not only body composition but also optimum slaughter weight can be manipulated. I speculate that β-agonists will have a limited role to play for conventional indoor strains of pigs, as they are partitioning agents that reduce fat deposition, thereby improving food conversion efficiency, without increasing lean deposition. In general, indoor pigs are already genetically sufficiently lean. By contrast, for outdoor-type pigs β-agonists may have a role to play.

Growth hormones appear to be an impressive way to manipulate not only body composition but also lean tissue growth rate, and therefore slaughter weight and the efficiency of the overall enterprise.

Welfare issues relating to an enterprise cannot be considered by a single criterion, be it behaviour, stress or thermal stress. The overall balance must be considered, and judgement is a personal issue. If one could develop a system that was a decided improvement compared to indoor units, with sows being kept outdoors and the piglets then brought indoors to finish on an *ad lib* system in straw yards, this would be likely to meet the welfare needs of the animals. At present, carcase quality in progeny on an *ad lib* feeding system would be adversely affected by the use of sows suited to the outdoor environment. However, it is envisaged that carcase composition could be manipulated hormonally to produce the required quality of product. But the crucial question is whether consumers would buy meat from pigs reared in this way.

Finally, there is an increasing gulf between consumer perception of what happens on a farm and the reality of agriculture. The only way to resolve this is to increase public knowledge by proper education, starting by schools bringing pupils on to the farm.

REFERENCES AND FURTHER READING

Clark, J.A. (1981) *Environmental Aspects of Housing for Animal Production.* Butterworths, London.

Monteith, J. and Mount, L.E. (Eds) (1974) *Heat Loss from Animals and Man: Its Assessment and Control.* Proceedings of the 20th Easter School

University of Nottingham School of Agriculture. Butterworths, London.

Webster, A.J.F. (1983) Nutrition and the thermal environment. In: J.A.F. Rook and P.C. Thomas (Eds). *Nutritional Physiology of Farm Animals*. Longman, London, 639-669.

Webster, A.J.F. (1987) Meat and right, farming as if the animal mattered. *Canadian Veterinary Journal*, **28**, 462-466.

DISCUSSION

Although insulation of arks has little effect on heat loss from well-bedded piglets in winter, it has a much greater influence on heat gain in summer and thus on the willingness of sows to enter their shelter and gain shade. Alternatively, modifications may be made to the nature of the roof surface to reduce solar gain and, particularly in tropical conditions, high rush sunshades which allow air movement may be more useful and effective than insulating the arks.

Most of the additional feed requirements of outdoor animals are due to cold conditions rather than to increased activity, which in turn leads to a higher heat production. In cold conditions, as feed energy is largely converted to heat, the total metabolisable energy intake of animals is much more important than the nutrient density of the feed. There would, however, be slight advantages to using feeds with a high heat increment, in terms of total heat production.

Welfare issues concerned with sow stalls are highly complex, the main problems being associated with tissue lesions and boredom, which can be overcome to a large extent by modifications to the conventional system. It is probable that the decision on whether to legislate against their use will be made on political grounds.